ASE Test Preparation

Medium/Heavy Duty Truck Technician Certification Series

Preventive Maintenance Inspection (T8)
5th Edition

DELMAR
CENGAGE Learning

Australia • Brazil • Japan • Korea • Mexico • Singapore • Spain • United Kingdom • United States

ASE Test Preparation: Medium/Heavy Duty Truck Technician Certification Series, Preventive Maintenance Inspection (T8), 5th Edition

Vice President, Technology and Trades Professional Business Unit: Gregory L. Clayton

Director, Professional Transportation Industry Training Solutions: Kristen L. Davis

Product Manager: Katie McGuire

Editorial Assistant: Danielle Filippone

Director of Marketing: Beth A. Lutz

Marketing Manager: Jennifer Barbic

Senior Production Director: Wendy Troeger

Production Manager: Sherondra Thedford

Content Project Management: PreMediaGlobal

Senior Art Director: Benjamin Gleeksman

Image Copyright Kirsty Pargeter, 2013. Used under license from Shutterstock.com

For product information and technology assistance, contact us at
Cengage Learning Customer & Sales Support, 1-800-354-9706
For permission to use material from this text or product,
submit all requests online at **www.cengage.com/permissions**
Further permissions questions can be emailed to
permissionrequest@cengage.com

ISBN-13: 978-1-111-12904-0

ISBN-10: 1-111-12904-5

Delmar
Executive Woods
5 Maxwell Drive
Clifton Park, NY 12065-2919
USA

Cengage Learning is a leading provider of customized learning solutions with office locations around the globe, including Singapore, the United Kingdom, Australia, Mexico, Brazil, and Japan. Locate your local office at **www.cengage.com/global**

Cengage Learning products are represented in Canada by Nelson Education, Ltd.

To learn more about Delmar, visit **www.cengage.com/delmar**

Purchase any of our products at your local bookstore or at our preferred online store **www.cengagebrain.com**

Notice to the Reader

Publisher does not warrant or guarantee any of the products described herein or perform any independent analysis in connection with any of the product information contained herein. Publisher does not assume, and expressly disclaims, any obligation to obtain and include information other than that provided to it by the manufacturer. The reader is expressly warned to consider and adopt all safety precautions that might be indicated by the activities described herein and to avoid all potential hazards. By following the instructions contained herein, the reader willingly assumes all risks in connection with such instructions. The publisher makes no representations or warranties of any kind, including but not limited to, the warranties of fitness for particular purpose or merchantability, nor are any such representations implied with respect to the material set forth herein, and the publisher takes no responsibility with respect to such material. The publisher shall not be liable for any special, consequential, or exemplary damages resulting, in whole or part, from the readers' use of, or reliance upon, this material.

Printed in the United States of America
2 3 4 5 6   26 25 24 23 22

Table of Contents

Delmar, a part of Cengage Learning, is very pleased that you have chosen to use our ASE Test Preparation Guide to help prepare yourself for the Preventive Maintenance Inspection (T8) ASE certification examination. This guide is designed to help prepare you for your actual exam by providing you with an overview and introduction of the testing process, introducing you to the task list for the Preventive Maintenance Inspection (T8) certification exam, giving you an understanding of what knowledge and skills you are expected to have in order to successfully perform the duties associated with each task area, and providing you with several preparation exams designed to emulate the live exam content in hopes of assessing your overall exam readiness.

If you have a basic working knowledge of the discipline you are testing for, you will find this book is an excellent guide, helping you understand the "must know" items needed to successfully pass the ASE certification exam. This manual is not a textbook. Its objective is to prepare the individual who has the existing requisite experience and knowledge to attempt the challenge of the ASE certification process. This guide cannot replace the hands-on experience and theoretical knowledge required by ASE to master the vehicle repair technology associated with this exam. If you are unable to understand more than a few of the preparation questions and their corresponding explanations in this book, it could be that you require either more shop-floor experience or further study.

This book begins by providing an overview of, and introduction to, the testing process. This section outlines what we recommend you do to prepare, what to expect on the actual test day, and overall methodologies for your success. This section is followed by a detailed overview of the ASE task list to include explanations of the knowledge and skills you must possess to successfully answer questions related to each particular task. After the task list, we provide six sample preparation exams for you to use as a means of evaluating areas of understanding, as well as areas requiring improvement in order to successfully pass the ASE exam. Delmar is the first and only test preparation organization to provide so many unique preparation exams. We enhanced our guides to include this support as a means of providing you with the best preparation product available. Section 6 of this guide includes the answer keys for each preparation exam, along with the answer explanations for each question. Each answer explanation also contains a reference back to the related task or tasks that it assesses. This will provide you with a quick and easy method for referring back to the task list whenever needed. The last section of this book contains blank answer sheet forms you can use as you attempt each preparation exam, along with a glossary of terms.

OUR COMMITMENT TO EXCELLENCE

Thank you for choosing Delmar, Cengage Learning for your ASE test preparation needs. All of the writers, editors, and Delmar staff have worked very hard to make this test preparation guide second to none. We feel confident that you will find this guide easy to use and extremely beneficial as you prepare for your actual ASE exam.

Delmar, Cengage Learning has sought out the best subject matter experts in the country to help with the development of *ASE Test Preparation: Medium/Heavy Duty Truck Technician Certification Series, Preventive Maintenance Inspection (T8), 5th Edition*. Preparation questions are authored and then reviewed by a group of certified, subject-matter experts to ensure the highest level of quality and validity to our product.

If you have any questions concerning this guide or any guide in this series, please visit us on the web at **http://www.trainingbay.cengage.com.**

For web-based online test preparation for ASE certifications, please visit us on the web at **http://www.techniciantestprep.com/** to learn more.

ABOUT THE AUTHOR

Doug Anderson has been involved in the truck and heavy equipment-related fields for over 30 years in various capacities. Doug is the Division Chair for Trades and Industry Programs at North Idaho College and was lead instructor of the Diesel Technology Program for 15 years. He has worked on many Cengage projects over the years and holds a Master's degree in Education and Bachelor of Science degree in Diesel Technology.

ABOUT THE SERIES ADVISOR

Brian (BJ) Crowley has experienced several different aspects of the diesel industry over the past 10 years. Now a diesel technician in the oil and gas industry, BJ owned and operated a diesel repair shop where he repaired heavy, medium, and light trucks—in addition to agricultural and construction equipment. He earned an Associate's degree in Diesel Technology from Elizabethtown Community and Technical College in Kentucky and is an ASE Master certified medium/heavy truck technician.

ASE began as the National Institute for Automotive Service Excellence (NIASE). It was founded as a non-profit, independent entity in 1972 by a group of industry leaders with the single goal of providing a means for consumers to distinguish between incompetent and competent technicians. It accomplishes this goal through the testing and certification of repair and service professionals. Though it is still known as the National Institute for Automotive Service Excellence, it is now called "ASE" for short.

Today, ASE offers more than 40 certification exams in automotive, medium/heavy duty truck, collision repair and refinish, school bus, transit bus, parts specialist, automobile service consultant, and other industry-related areas. At this time, there are more than 385,000 professionals nationwide with current ASE certifications. These professionals are employed by new car and truck dealerships, independent repair facilities, fleets, service stations, franchised service facilities, and more.

ASE's certification exams are industry-driven and cover practically every on-highway vehicle service segment. The exams are designed to stress the knowledge of job-related skills. Certification consists of passing at least one exam and documenting two years of relevant work experience. To maintain certification, those with ASE credentials must be re-tested every five years.

While ASE certifications are a targeted means of acknowledging the skills and abilities of an individual technician, ASE also has a program designed to provide recognition for highly qualified repair, support, and parts businesses. The Blue Seal of Excellence Recognition Program allows businesses to showcase their technicians and their commitment to excellence. One of the requirements of becoming Blue Seal recognized is that the facility must have a minimum of 75 percent of their technicians ASE certified. Additional criteria apply, and program details can be found on the ASE website.

ASE recognized that educational programs serving the service and repair industry also needed a way to be recognized as having the faculty, facilities, and equipment to provide a quality education to students wanting to become service professionals. Through the combined efforts of ASE, industry, and education leaders, the non-profit organization entitled the National Automotive Technicians Education Foundation (NATEF) was created in 1983 to evaluate and recognize academic programs. Today more than 2,000 educational programs are NATEF certified.

For additional information about ASE, NATEF, or any of their programs, the following contact information can be used:

National Institute for Automotive Service Excellence (ASE)

101 Blue Seal Drive S.E.

Suite 101

Leesburg, VA 20175

Telephone: 703-669-6600

Fax: 703-669-6123

Website: **www.ase.com**

Overview and Introduction

Participating in the National Institute for Automotive Service Excellence (ASE) voluntary certification program provides you with the opportunity to demonstrate you are a qualified and skilled professional technician who has the "know-how" required to successfully work on today's modern vehicles.

EXAM ADMINISTRATION

Note: After November 2011, ASE will no longer offer paper and pencil certification exams. There will be no Winter testing window in 2012, and ASE will offer and support CBT testing exclusively starting in April 2012.

CBT EXAMS

ASE provides computer-based testing (CBT) exams, which are administered at test centers across the nation. It is recommended that you go to the ASE website at **http://www.ase.com** and review the conditions and requirements for this type of exam. There is also an exam demonstration page that allows you to personally experience how this type of exam operates before you register.

CBT exams are available four times annually, for two-month windows, with a month of no testing in between each testing window:

- January/February – Winter testing window
- April/May – Spring testing window
- July/August – Summer testing window
- October/November – Fall testing window

Note: Testing windows and timing may change. It is recommended you go to the ASE website at **http://www.ase.com** and review the latest testing schedules.

UNDERSTANDING TEST QUESTION BASICS

ASE exam questions are written by service industry experts. Each question on an exam is created during an ASE-hosted "item-writing" workshop. During these workshops, expert service representatives from manufacturers (domestic and import), aftermarket parts and equipment manufacturers, working technicians, and technical educators gather to share ideas and convert them into actual exam questions. Each exam question written by these experts must then survive review by all members of the group. The questions are designed to address the practical application of repair and diagnosis knowledge and skills practiced by technicians in their day-to-day work.

After the item-writing workshop, all questions are pre-tested and quality-checked on a national sample of technicians. Those questions that meet ASE standards of quality and accuracy are included in the scored sections of the exams; the "rejects" are sent back to the drawing board or discarded altogether.

Depending on the topic of the certification exam, you will be asked between 40 and 80 multiple-choice questions. You can determine the approximate number of questions you can expect to be asked during the Preventive Maintenance Inspection (T8) certification exam by reviewing the task list in Section 4 of this book. The five-year recertification exam will cover this same content; however, the number of questions for each content area of the recertification exam will be reduced by approximately one-half.

> *Note:* Exams may contain questions that are included for statistical research purposes only. Your answers to these questions will not affect your score, but since you do not know which ones they are, you should answer all questions in the exam.

Using multiple criteria, including cross-sections by age, race, and other background information, ASE is able to guarantee that exam questions do not include bias for or against any particular group. A question that shows bias toward any particular group is discarded.

TEST-TAKING STRATEGIES

Before beginning your exam, quickly look over the exam to determine the total number of questions that you will need to answer. Having this knowledge will help you manage your time throughout the exam to ensure you have enough available to answer all of the questions presented. Read through each question completely before marking your answer. Answer the questions in the order they appear on the exam. Leave the questions blank that you are not sure of and move on to the next question. You can return to those unanswered questions after you have finished the others. These questions may actually be easier to answer at a later time once your mind has had additional time to consider them on a subconscious level. In addition, you might find information in other questions that will help you recall the answers to some of them.

Multiple-choice exams are sometimes challenging because there are often several choices that may seem possible, or partially correct, and therefore it may be difficult to decide on the most appropriate answer choice. The best strategy, in this case, is to first determine the correct answer before looking at the answer options. If you see the answer you decided on, you should still be careful to examine the other answer options to make sure that none seems more correct than yours. If you do not know or are not sure of the answer, read each option very carefully and try to eliminate those options that you know are incorrect. That way, you can often arrive at the correct choice through a process of elimination.

If you have gone through the entire exam and you still do not know the answer to some of the questions, *then guess*. Yes, guess. You then have at least a 25 percent chance of being correct. While your score is based on the number of questions answered correctly, any question left blank, or unanswered, is automatically scored as incorrect.

There is a lot of "folk" wisdom on the subject of test taking that you may hear about as you prepare for your ASE exam. For example, there are those who would advise you to avoid response options that use certain words such as *all, none, always, never, must,* and *only,* to name a few. This, they claim, is because nothing in life is exclusive. They would advise you to choose response options that use words that allow for some exception, such as *sometimes, frequently, rarely, often, usually, seldom,* and *normally.* They would also advise you to avoid the first and last option (A or D) because exam writers, they feel, are more comfortable if they put the correct answer in the middle (B or C) of the choices. Another recommendation often offered is to select the option that is either shorter or longer than the other three choices because it is more likely to be correct. Some would advise you to never change an answer since your first intuition is usually correct. Another area of "folk" wisdom focuses specifically on any repetitive patterns created by your question responses (e.g., A, B, C, A, B, C, A, B, C).

Many individuals may say that there are actual grains of truth in this "folk" wisdom, and whereas with some exams, this may prove true, it is not relevant in regard to the ASE certification exams. ASE validates all exam questions and test forms through a national sample of technicians, and only those questions and test forms that meet ASE standards of quality and accuracy are included in the scored sections of the exams. Any biased questions or patterns are discarded altogether, and therefore, it is highly unlikely you will experience any of this "folk" wisdom on an actual ASE exam.

PREPARING FOR THE EXAM

Delmar, Cengage Learning wants to make sure we are providing you with the most thorough preparation guide possible. To demonstrate this, we have included hundreds of preparation questions in this guide. These questions are designed to provide as many opportunities as possible to prepare you to successfully pass your ASE exam. The preparation approach we recommend and outline in this book is designed to help you build confidence in demonstrating what task area content you already know well, while also outlining what areas you should review in more detail prior to the actual exam.

We recommend that your first step in the preparation process should be to thoroughly review Section 3 of this book. This section contains a description and explanation of the type of questions you'll find on an ASE exam.

Once you understand how the questions will be presented, we then recommend that you thoroughly review Section 4 of this book. This section contains information that will help you establish an understanding of what the exam will be evaluating, and specifically, how many questions to expect in each specific task area.

As your third preparatory step, we recommend you complete your first preparation exam, located in Section 5 of this book. Answer one question at a time. After you answer each question, review the answer and question explanation information located in Section 6. This section will provide you with instant response feedback, allowing you to gauge your progress, one question at a time, throughout this first preparation exam. If after reading the question explanation you don't feel you understand the reasoning for the correct answer, go back and review the task list overview (Section 4) for the task that is related to that question. Included with each question explanation is a clear identifier of the task area that is being assessed (e.g., Task A.1). If at that point you still don't feel you have a solid understanding of the material, identify a good source of information on the topic, such as an educational course, textbook, or other related source of topical learning, and do some additional studying.

After you have completed your first preparation exam and have reviewed your answers, you are ready to complete your next preparation exam. A total of six practice exams are available in Section 5 of this book. For your second preparation exam, we recommend that you answer the questions as if you were taking the actual exam. Do not use any reference material or allow any interruptions in order to get a feel for how you will do on the actual exam. Once you have answered all of the questions, grade your results using the Answer Key in Section 6. For every question that you gave an incorrect answer to, study the explanations to the answers and/or the overview of the related task areas. Try to determine the root cause for missing the question. The easiest thing to correct is learning the correct technical content. The hardest things to correct are behaviors that lead you to an incorrect conclusion. If you knew the information but still got the question incorrect, there is likely a test-taking behavior that will need to be corrected. An example of this would be reading too quickly and skipping over words that affect your reasoning. If you can identify what you did that caused you to answer the question incorrectly, you can eliminate that cause and improve your score.

Here are some basic guidelines to follow while preparing for the exam:

- Focus your studies on those areas you are weak in.
- Be honest with yourself when determining if you understand something.
- Study often but for short periods of time.
- Remove yourself from all distractions when studying.
- Keep in mind that the goal of studying is not just to pass the exam; the real goal is to learn.
- Prepare physically by getting a good night's rest before the exam, and eat meals that provide energy but do not cause discomfort.
- Arrive early to the exam site to avoid long waits as test candidates check in.
- Use all of the time available for your exams. If you finish early, spend the remaining time reviewing your answers.
- Do not leave any questions unanswered. If absolutely necessary, guess. All unanswered questions are automatically scored as incorrect.

Here are some items you will need to bring with you to the exam site:

- A valid government or school-issued photo ID
- Your test center admissions ticket
- A watch (not all test sites have clocks)

> *Note:* Books, calculators, and other reference materials are not allowed in the exam room. The exceptions to this list are English-Foreign dictionaries or glossaries. All items will be inspected before and after testing.

WHAT TO EXPECT DURING THE EXAM

When taking a CBT exam, as soon as you are seated in the testing center, you will be given a brief tutorial to acquaint you with the computer-delivered test, prior to taking your certification exam(s). The CBT exams allow you to select only one answer per question. You can also change your answers as many times as you like. When you select a second answer choice, the CBT will automatically unselect your first answer choice. If you want to skip a question to return to later, you can utilize the "flag" feature, which will allow you to quickly identify and review questions whenever you are ready. Prior to completing your exam, you will also be provided with an opportunity to review your answers and address any unanswered questions.

TESTING TIME

Each individual ASE CBT exam has a fixed time limit. Individual exam times will vary based upon exam area, and will range anywhere from a half hour to two hours. You will also be given an additional 30 minutes beyond what is allotted to complete your exams to ensure you have adequate time to perform all necessary check-in procedures, complete a brief CBT tutorial, and potentially, complete a post-test survey.

You can register for and attempt multiple CBT exams during one testing appointment. The maximum time allotment for a CBT appointment is four and a half hours. If you happen to register for so many exams that you will require more time than this, your exams will be scheduled into multiple appointments. This could mean that you have testing on both the morning and afternoon of the same day, or they could be scheduled on different days, depending on your personal preference and the test center's schedule.

It is important to understand that if you arrive late for your CBT test appointment, you will not be able to make up any missed time. You will only have the scheduled amount of time remaining in your appointment to complete your exam(s).

Also, while most people finish their CBT exams within the time allowed, others might feel rushed or not be able to finish the test, due to the implied stress of a specific, individual time limit allotment. Before you register for the CBT exams, you should review the number of exam questions that will be asked along with the amount of time allotted for that exam to determine whether you feel comfortable with the designated time limitation or not.

Summary

As an overall time management recommendation, you should monitor your progress and set a time limit you will follow with regard to how much time you will spend on each individual exam question. This should be based on the total number of questions you are attempting.

Also, it is very important to note that if for any reason you wish to leave the testing room during an exam, you must first ask permission. If you happen to finish your exam(s) early and wish to leave the testing site before your designated session appointment is completed, you are permitted to do so only during specified dismissal periods.

UNDERSTANDING HOW YOUR EXAM IS SCORED

You can gain a better perspective about the ASE certification exams if you understand how they are scored. ASE exams are scored by an independent organization having no vested interest in ASE or in the automotive industry. With CBT exams, you will receive your exam scores immediately.

Each question carries the same weight as any other question. For example, if there are 50 questions, each is worth 2 percent of the total score. The passing grade is 70 percent.

Your exam results can tell you

- Where your knowledge equals or exceeds that needed for competent performance, or
- Where you might need more preparation.

Your ASE exam score report is divided into content "task" areas; it will show the number of questions in each content area and how many of your answers were correct. These numbers provide information about your performance in each area of the exam. However, because there may be a different number of questions in each content area of the exam, a high percentage of correct answers in an area with few questions may not offset a low percentage in an area with many questions.

It should be noted that one does not "fail" an ASE exam. The technician who does not pass is simply told "More Preparation Needed." Though large differences in percentages may indicate problem areas, it is important to consider how many questions were asked in each area. Since each exam evaluates all phases of the work involved in a service specialty, you should be prepared in each area.

A low score in one area could keep you from passing an entire exam. If you do not pass the exam, you may take it again at any time it is scheduled to be administered.

There is no such thing as average. You cannot determine your overall exam score by adding the percentages given for each task area and dividing by the number of areas. It doesn't work that way because there generally is not the same number of questions in each task area. A task area with 20 questions, for example, counts more toward your total score than a task area with 10 questions.

Your exam report should give you a good picture of your results and a better understanding of your strengths and areas needing improvement for each task area.

Understanding not only what content areas will be assessed during your exam, but how you can expect exam questions to be presented will enable you to gain the confidence you need to successfully pass an ASE certification exam. The following examples will help you recognize the types of question styles used in ASE exams and assist you in avoiding common errors when answering them.

Most initial certification tests are made up of between 40 and 80 multiple-choice questions. The five-year recertification exams will cover the same content as the initial exam; however, the actual number of questions for each content area will be reduced by approximately one-half. Refer to Section 4 of this book for specific details regarding the number of questions to expect during the initial Preventive Maintenance Inspection (T8) certification exam.

Multiple-choice questions are an efficient way to test knowledge. To correctly answer them, you must consider each answer choice as a possibility, and then choose the answer choice that *best* addresses the question. To do this, read each word of the question carefully. Do not assume you know what the question is asking until you have finished reading the entire question.

About 10 percent of the questions on an actual ASE exam will reference an illustration. These drawings contain the information needed to correctly answer the question. The illustration should be studied carefully before attempting to answer the question. When the illustration is showing a system in detail, look over the system and try to figure out how the system works before you look at the question and the possible answers. This approach will ensure that you do not answer the question based upon false assumptions or partial data, but instead have reviewed the entire scenario being presented.

MULTIPLE-CHOICE/DIRECT QUESTIONS

The most common type of question used on an ASE exam is the direct multiple-choice style question. This type of question contains an introductory statement, called a stem, followed by four options: three incorrect answers, called distracters, and one correct answer, the key.

When the questions are written, the point is to make the distracters plausible to draw an inexperienced technician to inadvertently select one of them. This type of question gives a clear indication of the technician's knowledge.

Here is an example of a direct multiple-choice style question:

TASK E.6

1. An electronic cruise control maintains the vehicle speed at 45 mph (72 km/h) when the selected speed is 60 mph (96 km/h). What is the most likely cause of this problem?

    A. A defective cruise control module

    B. A defective vehicle speed sensor (VSS)

    C. Improper size of tires on the drive axles

    D. A defective cruise control switch

Answer A is incorrect. A defective cruise control module is not the most likely cause of improper cruise set speed.

Answer B is correct. A VSS is the most likely cause of improper cruise control set speed.

Answer C is incorrect. Improper size of tires on the drive axles is not the most likely cause of improper cruise control set speed.

Answer D is incorrect. A defective cruise control switch is not the most likely cause of improper cruise control set speed.

COMPLETION QUESTIONS

A completion question is similar to the direct question except the statement may be completed by any one of the four options to form a complete sentence. Here's an example of a completion question:

1. When aiming headlights with a headlight aligning screen:

    A. The vehicle does not need to be parallel to the aligning screen.
    B. The tire inflation pressure does not affect headlight alignment.
    C. The headlight alignment may be changed by installing new sealed beams.
    D. The headlight alignment is changed by turning adjusting screws on the headlight assembly.

TASK C.11

Answer A is incorrect. The vehicle does need to be parallel to the aligning screen.

Answer B is incorrect. The tire inflation pressure does affect headlight alignment.

Answer C is incorrect. The headlight alignment is not changed by installing new sealed beams.

Answer D is correct. The headlight alignment is changed by turning adjusting screws on the headlight assembly.

TECHNICIAN A, TECHNICIAN B QUESTIONS

This type of question is most popularly associated with an ASE exam. It is, in fact, two true-false statements grouped together, such as: "Technician A says..." and "Technician B says..."..." Who is correct?"

In this type of question, you must determine whether either, both, or neither of the statements are correct. To answer this type of question correctly, you must carefully read each technician's statement and judge it on its own merit.

Sometimes this type of question begins with a statement about an inspection or maintenance procedure. This statement provides the setup or background information required to understand the conditions about which Technician A and Technician B are talking, followed by two statements about the cause of the concern, proper inspection, identification, or repair choices.

Analyzing this type of question is a little easier than the other question types because there are only two ideas to consider, although there are still four choices for an answer.

Again, Technician A, Technician B questions are really double true-or-false statements. The best way to analyze this type of question is to consider each technician's statement separately. Ask yourself, "Is A true or false? Is B true or false?" Once you have completed an individual evaluation of each statement, you will have successfully determined the correct answer choice.

An important point to remember is that an ASE Technician A, Technician B question will never have Technician A and B directly disagreeing with each other. That is why you must evaluate each statement independently.

An example of a Technician A/Technician B style question looks like this:

TASK A.14

1. During a PMI, Technician A says the air passages through the air-to-air intercooler should be checked for restrictions. Technician B says the air-to-air intercooler uses refrigerant to lower the intake air temperature. Who is correct?

    A. A only

    B. B only

    C. Both A and B

    D. Neither A nor B

Answer A is correct. Only Technician A is correct. The air passages through the air-to-air intercooler should be checked for restrictions.

Answer B is incorrect. The air-to-air intercooler does not use refrigerant to lower intake air temperature.

Answer C is incorrect. Only Technician A is correct.

Answer D is incorrect. Technician A is correct.

EXCEPT QUESTIONS

Another type of question used on ASE exams contains answer choices that are all correct except for one. To help easily identify this type of question, whenever they are presented in an exam, the word "EXCEPT" will always be displayed in capital letters. Furthermore, a cautionary statement will alert you to the fact that the next question is different from the questions otherwise found in the exam. With the EXCEPT type of question, only one *incorrect* choice will actually be listed among the options, and that incorrect choice will be the key to the question. That is, the incorrect statement is counted as the correct answer for that question.

Be careful to read these question types slowly and thoroughly; otherwise, you may overlook what the question is actually asking and answer the question by selecting the first correct statement.

An example of this type of question would appear as follows:

TASK A.15

1. A turbocharger on a tractor has less than the specified boost pressure. All of these defects may be the cause of the problem EXCEPT:

    A. Turbocharger bearings that are starting to fail.

    B. Improper wastegate adjustment.

    C. Damaged compressor wheel.

    D. Turbocharger oil seals leaking oil into the exhaust.

Answer A is incorrect. Turbocharger bearings that are starting to fail may cause reduced boost pressure.

Answer B is incorrect. Improper wastegate adjustment may cause reduced boost pressure.

Answer C is incorrect. A damaged compressor wheel may cause reduced boost pressure.

Answer D is correct. Turbocharger oil seals leaking oil into the exhaust cause blue smoke in the exhaust, but this problem does not cause reduced boost pressure.

LEAST LIKELY QUESTIONS

LEAST LIKELY questions are similar to EXCEPT questions. Look for the answer choice that would be the LEAST LIKELY cause (most incorrect) for the described situation. To help easily identify these types of questions, whenever they are presented in an exam the words "LEAST LIKELY" will always be displayed in capital letters. In addition, you will be alerted before a LEAST LIKELY question is posed. Read the entire question carefully before choosing your answer.

An example of this type of question is shown below:

TASK D.1.15

1. A truck with an air brake system has reduced braking force on one rear wheel. Which of the following is the LEAST LIKELY cause?

    A.   Restricted airflow or low air pressure to the service brake chamber
    B.   Improper adjustment of slack adjuster and brake chamber pushrod
    C.   Grease on shoe linings/faces
    D.   A ruptured brake chamber diaphragm

Answer A is incorrect. A condition causing a lowering of air pressure will subsequently reduce service brake application force.

Answer B is incorrect. An incorrect or lack of adjustment to a slack adjuster will cause a reduction in overall brake application forces.

Answer C is correct. Grease on shoe linings/faces causes grabbing. It is the least likely cause of insufficient braking.

Answer D is incorrect. A ruptured brake chamber diaphragm will severely decrease an air brake system's service application force.

SUMMARY

The question styles outlined above are the only ones you will encounter on any ASE certification exam. ASE does not use any other types of question styles, such as fill-in-the-blank, true/false, word-matching, or essay. ASE also will not require you to draw diagrams or sketches to support any of your answer selections; although any of the above described question styles may include illustrations, charts, or schematics to clarify a question. If a formula or chart is required to answer a question, it will be provided for you. You may, in rare cases, be required to solve a simple math problem, so bringing a simple pocket calculator to the test session might be a good idea.

INTRODUCTION

This section of the book outlines the content areas, or *task list,* for this specific certification exam, along with a written overview of the content covered in the exam.

The task list describes the actual knowledge and skills necessary for a technician to successfully perform the work associated with each skill area. This task list is the fundamental guideline you should use to understand what areas you can expect to be tested on, as well as how each individual area is weighted to include the approximate number of questions you can expect to be given for that area during the ASE certification exam. It is important to note that the number of exam questions for a particular area is to be used as a guideline only. ASE advises that the questions on the exam may not equal the number specifically listed on the task list. The task lists are specifically designed to tell you what ASE expects you to know how to do and to help you prepare to be tested.

Similar to the role this task list will play with regard to the actual ASE exam, Delmar, Cengage Learning has developed six preparation exams, located in Section 5 of this book, using this task list as a guide. It is important to note that although both ASE and Delmar, Cengage Learning use the same task list as a guideline for creating these test questions, none of the test questions you will see in this book will be found in the actual, live ASE exams. This is true for any test preparatory material you use. Real exam questions are *only* visible during the actual ASE exams.

Task List at a Glance

The Preventive Maintenance Inspection (T8) task list focuses on five core areas, and you can expect to be asked a total of approximately 50 questions on your certification exam, broken out as outlined here:

- A. Engine Systems (10 Questions)
- B. Cab and Hood (5 Questions)
- C. Electrical/Electronics (10 Questions)
- D. Frame and Chassis (22 Questions)
    1. Brakes (10)
    2. Drivetrain (4)
    3. Steering, Tires, and Wheels (4)
    4. Suspension, Frame, and 5th Wheel (4)
- E. Road/Operational Test (3 Questions)

Based upon this information, the graph shown here is a general guideline demonstrating which areas will have the most focus on the actual certification exam. This data may help you prioritize your time when preparing for the exam.

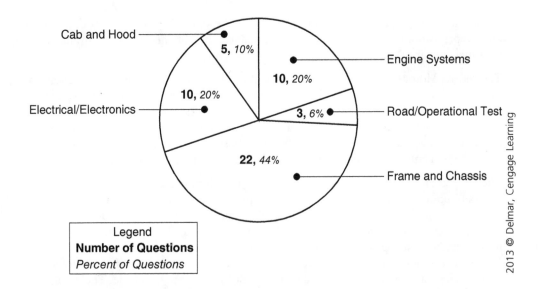

Note: The actual number of questions you will be given on the ASE certification exam may vary slightly from the information provided in the task list, as exams may contain questions that are included for statistical research purposes only. Do not forget that your answers to these research questions will not affect your score.

PREVENTIVE MAINTENANCE INSPECTION (TEST T8) TASK LIST

A. Engine Systems (10 Questions)

1. Review past maintenance/repair documents, driver vehicle condition reports, and brake/tire wear report. Check and record electronic diagnostic codes and trip/operational data. Clear codes and data.

The decision to take a vehicle out of service (OOS) can be a tough one. It is necessary to evaluate whether the mechanical problem will cause an accident or breakdown versus the need to get the load moved. Management can minimize deadline situations by listening to the driver and technician and making decisions based on the given facts and possible consequences. The driver (and technician) should perform a pre-trip inspection. The inspection should start at the left front of the vehicle and work in a counterclockwise direction around it. Permanent records are invaluable from a performance standpoint. A study of failures of individual components can alert the manager of the need to take corrective action. It also highlights components that have performed satisfactorily.

Whenever any repair procedure is performed on a vehicle, the work must be noted using a sign-off sheet before the vehicle is put back into service. If the problem noted did not require repair, but was still looked at, then this must be recorded as well. A copy of the inspection report must remain with the vehicle.

Service logs should reflect work performed on a vehicle. Active complaints/problems should always be addressed immediately with a quick visual inspection to determine if the repair can be made quickly and easily. Then check the vehicle's records if it appears that a quick repair cannot be made. All information that can be obtained could be invaluable to the timely repair of the vehicle.

The Federal Motor Carrier Safety Regulations, Part 393, states that anything mechanical that can either cause or prevent an accident is a safety item. All vehicle safety problems should be addressed, no matter how minor they appear. Section 396.19 of the Federal Motor Carrier Safety Regulations states it is the fleet manager's responsibility to ensure that all personnel doing annual inspections are qualified to perform those inspections. They should understand the inspection criteria and be able to identify defective components and systems.

The Federal Highway Administration has set up a minimum inspection standards program, under which each vehicle must carry proof that an inspection was completed. Proof can either be a copy of the inspection form kept on the vehicle or a decal. If using a decal, a copy of the inspection form must be kept on file and should indicate where an inspector can get a copy of it. The only requirement of a decal itself is that it be legible. Each vehicle must be inspected separately. This means a tractor/trailer is considered to be two vehicles, each requiring a decal; a converter dolly is a separate vehicle. The decal must show the following information: date vehicle passed inspection, name and address to contact concerning inspection record, and vehicle identification number (VIN).

Although the regulations seem to place greater emphasis on the post-trip inspection, most operators agree the pre-trip inspection is more important. Because the driver doing the inspection will be operating that particular vehicle, the incentive to ensure that the vehicle is safe is probably greater. Repairs are cheaper and less time-consuming if the driver finds them on the pre-trip before inspectors levy fines or impose penalties.

Although still a part of the overall trailer assembly, a refrigeration unit, or "reefer" as it is generally called, would not be considered an underside component of a trailer. Reefer units are commonly mounted on the upper portion of the trailer bulkhead where they can perform most efficiently.

Today's diesel trucks contain a variety of computerized control systems to operate engines, transmissions, antilock brakes, etc. According to Society of Automotive Engineers (SAE) Standard J1930, diagnostic codes are known as diagnostic trouble codes (DTCs). DTC access may be by blink codes or read off the chassis databus. You need to understand how to check, record, and clear DTCs. Task B.7 provides more information on DTCs.

2. Check engine operation (including unusual noises, vibration, and excessive exhaust smoke); record idle rpm, governed rpm, and PTO rpm (if applicable).

The engine should crank when the starter is engaged at a sufficient revolutions per minute (RPM). See manufacturers' specifications because they differ regarding RPM. Electronic diesel and gasoline-fueled engines will not inject fuel until at least 150 rpm. Using ether can cause detonation (loud hammering sound). It should never be used on glow plug applications.

White exhaust smoke is an indication of unburned fuel exiting the exhaust pipe. Blue smoke is an indication of burning oil. Black smoke is caused by air starvation or a rich-fuel mixture. Governed RPM is the maximum engine RPM that an engine is rated for and the fuel or engine-management system will attempt to limit speeds above that setting.

3. Inspect vibration damper.

As part of the preventive maintenance inspection (PMI), the engine is inspected, including the vibration damper. This vital engine component is mounted on the front of the engine, balancing harmonic vibrations caused by torsional loads of the rotating crankshaft. The damper is usually inspected for leaks and cracks caused by stress, dents, abrasions, and/or deterioration. A loose, damaged, or defective vibration damper can cause crankshaft failure.

4. Inspect condition of belt(s), tensioner(s), and pulley(s); check and adjust belt tension.

All accessory belts should be checked for obvious signs of wear, tension, and/or correct application (if suspected). A belt tension gauge (shown in the following illustration) should be used whenever possible for the most accurate readings. If one is not available, then the normal deflection points at the specified locations and measurements should be observed. An idler hub should be checked for loose or worn bearings, condition of belt surfaces, cracks from fatigue or deterioration, and the correct application.

5. Check engine for oil, coolant, air, and fuel leaks (Engine Off and Engine Running).

An engine should be checked while doing an inspection for both static and dynamic leaks. This means that the engine should be checked while it is running and while it is off. This ensures that all gasket and mating surfaces are monitored in all possible operating and non-operating conditions. When checking for fuel leaks on an engine, it is mandatory that the engine be operating so the fuel system is pressurized, thus making a leak very apparent. A fuel leak while the engine is running is more hazardous than when the engine is off due to the higher fuel pressure with the engine running. An inspection of the cooling system should include a check of the cooling system maintenance history, coolant change intervals, component replacement(s), and past diagnoses. Although a leak in the cooling system can be a serious problem, it is not criteria for deadlining because it is not safety related.

6. Inspect engine mounts for looseness and deterioration.

The most effective way of diagnosing a vehicle with defective engine mounts is to observe the engine mounts while torquing the engine with the brakes applied. This will allow a broken or deteriorated engine mount to be quickly observed and noted for replacement.

7. Check engine oil level and condition; check dipstick seal and fill cap seal.

Checking the engine oil level is a critical part of a PMI. The correct oil level is essential for proper engine lubrication and cooling. An oil level that is too low may allow air to be drawn into the pump and a loss of oil circulation, resulting in engine failure. An oil level that is too high may allow the oil to be whipped by the crankshaft, which could cause oil aeration. Aerated oil is compressible and does not provide a continuous flow of oil to the various components that depend upon oil for lubrication, cooling, or hydraulic operation. This condition could result in engine failure or accelerated wear. The recommended procedure for checking the oil level is to ensure that the vehicle is on level ground and has not been operated for at least five minutes. The dipstick must be removed and wiped with a clean rag to remove any oil from previous operation. Fully insert the clean dipstick and remove to obtain the correct reading.

8. Check engine compartment wiring, harnesses, connectors, and seals for damage, mounting, and proper routing.

Because so many systems on a modern truck depend upon the electrical circuit in one way or another, careful and frequent examination of the wiring harnesses is essential. Pay particular attention to any areas where the wiring harness contacts anything metal. Wiring harnesses should be covered with protective armor and secured in place with brackets and loom holders. Make sure all wiring is properly routed to avoid damage from moving parts and hot exhaust manifolds. Modern connectors are constructed with weatherproof seals to prevent corrosion. These seals should always be in good condition because many electrical faults are traced to problems at the connectors. When it is necessary to remove a wire from a connector for repair or replacement, be sure to use the proper tools and replacement parts. Wherever a splice or other repair is made to the wiring, it should be sealed to weatherproof it.

9. Check fuel tanks, mountings, lines, caps, and seals; check anti-siphon device (if applicable).

The condition of the fuel tanks, mountings, lines, and caps are of concern during a PMI. Tanks must be checked for any signs of leakage at any point, with special attention to all seams and any damaged areas. The fuel tank mounts should be securely attached to the frame and the tank and should be free from any cracks, corrosion, and damage that could cause future failure. There should be no sign of tank shifting and all mounting bolts should be in place and tight. The tank fuel supply can also be checked for moisture contamination by the use of a rod or probe coated with water detection paste inserted into the tank. The paste is designed to change color when it comes into contact with water. If more than the tip changes color, the tanks may require draining to remove the water. Water in the fuel tank can be drained by the use of a suction pump with a long sump that reaches to the bottom of the tank where the water would accumulate, or by a tank drain if it is so equipped.

The fuel tank cap should be easily installed and removed and should be able to provide a good seal. Check to ensure that the cap seal is in place and in good condition. The fuel lines including the crossover line must be leak-free and should be secured by loom or brackets. Inspect all routing to ensure that no line damage will occur due to contact with other objects. If a line is chafed but no leak is present, a replacement decision should be made depending on the severity and the probability of a future leak or breakdown.

10. Inspect water separator/fuel heater; drain water from separator; replace fuel filter(s); prime and bleed fuel system.

When a combination water separator/primary filter is used, routine water draining and changing the filter element when the other engine filters are changed is a good practice. Some aftermarket filter elements have a longer service life and should be changed at the specified interval. Fuel contamination can occur easily if the canister is primed with unfiltered fuel or dirt is allowed to enter when the element is being changed. Most fuel filters are changed at routine intervals (PMI schedules). These filters must be primed with fuel before the engine can be started. Most fuel filters can be filled with clean fuel and reinstalled, but some manufacturers recommend that the secondary filter be installed dry and primed with a hand pump. This procedure ensures that only filtered diesel fuel enters the secondary filter. When hand priming is necessary, a line fitting upstream from the pump should be cracked open and priming should continue until no air bubbles are present at the fitting.

11. Inspect crankcase ventilation system.

The PCV valve should always be checked when any inspection of the emission control system is performed. Improper operation of the PCV can cause erratic engine operation as well as higher than normal gasoline exhaust emissions, thereby possibly causing fines and/or penalties due to violations.

12. Check exhaust system mountings for looseness and damage.

To check exhaust pipes and support brackets, a pry bar can be used to check for looseness. All mounting brackets should be examined for cracks and clamps must be tight. The complete system should be visually inspected for rust, damage, and cracks.

13. Check engine exhaust system for leaks, excessive noise, proper routing, and missing or damaged components (heat shields and guards).

Any exhaust system that is leaking in front of or below the driver/sleeper compartment and the floor pan has potential for cab entry of exhaust fumes. These fumes could result in a drowsy or asphyxiated driver, which could cause an accident. The exhaust routing must be checked and assembled in a way that does not allow contact of the exhaust system with any wiring, air lines, fuel lines, or combustible materials. The Commercial Vehicle Safety Alliance (CVSA) out-of-service criteria would deadline a vehicle for any of these faults.

14. Check air induction system piping, charge air cooler, hoses, clamps, mountings, and indicators; check for air restrictions and leaks.

Depending on the conditions the vehicle is driven in, truck air induction ducts do not generally have to be cleaned out on every PMI. It is always a good idea to give a full visual inspection to the air induction system for blockage and/or deterioration whenever a PMI is performed, only cleaning it when necessary. The air inlet restriction should be checked with a water manometer or accurate vacuum gauge. If the restriction exceeds the maximum specification, the air filter should be replaced.

Restricted dry filters should not be tapped on the floor or blown with air pressure to dislodge dirt. Both of these practices can result in filter damage and dirt bypassing the filter if it is reinstalled.

15. Inspect turbocharger(s) for noise, oil, and exhaust leaks; check mountings and connections, check wastegate, variable geometry turbocharger (VGT), linkages, and hose.

A turbocharger operates on the principle of thermal expansion to drive a turbine wheel, located in the turbine housing. This in turn spins the compressor wheel in the impeller housing. The compressor wheel is located in the inlet side and draws in fresh air and pressurizes it. It is necessary to contain this pressurized air in hoses, manifolds, and an intercooler. A turbocharger can operate within manufacturer's specifications and still have acceptable leakage. To test for proper leak rate, follow the most current information provided in manufacturers' published service bulletins and do not exceed the recommended test air pressure. The clamps used are spring loaded and should be torqued to specifications and placed in a way as to not chafe through other hoses or components with engine movement. Modern diesel engine turbochargers perform at very high speeds and can reach well over 100,000 rpm. When running the engine without the air intake, be sure to install an intake guard over the turbocharger inlet.

16. Service or replace air filter(s) as needed.

Engines require approximately 9,000 to 10,000 gallons of air by volume for every gallon of fuel consumed. This results in a volumetric air-fuel ratio of approximately 10,000 to 1 for both gasoline and diesel engines. Air-fuel ratio by mass is approximately 15:1 and is referred to as the stoichiometric ratio.

With this large volume of air entering an engine it is important to provide good filtration, as airborne dust, sand, and other debris ingested into the cylinders will lead to premature engine failure (called "dusting" caused by loose clamps or ill-fitting air filter assemblies). Obstructions such as a restricted air filter will reduce airflow/volume and cause an engine to operate in a rich (fuel) condition and cause cylinder wash down of the oil film on the cylinder walls, again leading to engine failure. Always service following manufacturer's specifications and intervals. Typical restriction specifications are 15 inches of water vacuum for naturally aspirated engines and 25 inches of water vacuum for boosted engines.

17. Inspect diesel emission control systems, including exhaust gas recirculation (EGR), exhaust gas recirculation (EGR) cooler, diesel particulate filter (DPF), and/or catalytic converter.

EGR is a method of reducing oxides of nitrogen (NOx) by circulating a small amount of exhaust gases back through the intake manifold to dilute the intake charge and lower the combustion temperature. The EGR system should only function on a warm engine, operating above idle. A common test of a vacuum-operated EGR valve is to apply vacuum to the valves diaphragm at idle, which should produce a rough idle or even stall out. At above-idle operation, vacuum should be present at the valve. Air inlet temperature control is simply a dual position door located in the air cleaner inlet, which allows direct inlet air in when the engine is warm and preheated air in when the engine is cold. The preheated air is air drawn past the exhaust manifolds through ducting and shrouds. Early fuel evaporation (EFE) is a process of diverting exhaust gases through a passage in the intake manifold during cold engine operation, to prevent fuel condensation on the cold walls. Some EFE systems use an electric heating grid to warm the intake. Evaporative emission (EVAP) systems use a sealed fuel tank with pressure relief vents that are plumbed to a carbon canister. The carbon canister absorbs the carbon from the gasoline vapor. When the engine is running, manifold vacuum draws the gasoline vapor into the engine to be burned. Catalytic converters are used on most gasoline engines. Catalytic converters use catalyst substances to convert harmful emissions into non-harmful emissions. Palladium and platinum are oxidation catalysts that attempt to oxidize hydrocarbons and carbon monoxide into water and carbon dioxide. Rhodium is a reduction catalyst that tries to reduce oxides of nitrogen into the elements nitrogen and oxygen.

Generally, diesel engines have used less external emission control devices. Computer-controlled diesels have very precise control over the combustion process, so most engines meet emission standards. An increasing number of diesel engines are being equipped with cooled-exhaust gas recirculation (C-EGR) to meet more stringent emission standards. Single stage or oxidizing catalytic converters are used on some engines while others use particulate traps to filter soot particles. To reduce cold engine emissions, some engines use intake air heaters to reduce unburned hydrocarbons.

Catalytic converter operation can be checked with an infrared thermometer. A correctly operating converter should operate in the range of 1,500 to 1,800°F and the inlet temperature should be less than the output temperature. All of the above systems can be checked and serviced relatively easily, but with such a broad range of systems, manufacturers' manuals should be consulted for correct service procedures.

18. Check fan clutch/hub operation, bearing condition and noise (including viscous/thermostatic, air, and electric fan types); inspect fan assembly and shroud for missing and damaged components.

Fan clutch types may be viscous, thermatic, air apply, air release, electric, and hydraulic. Each type has different characteristics and should first be identified to avoid improper diagnosis. Viscous and hydraulic units should not exhibit any signs of fluid leakage. Most pneumatic fan clutch failures are caused by air leaks. Whatever preventive maintenance (PM) level is being performed, an air leak check is critical to maintaining proper operation of the pneumatic fan clutch. The engine computer may control on-off operation. Fan clutches may be engaged by engine compartment temperatures, fluid temperatures, or by engine management strategies.

The fan is subject to receiving road debris created by the ram air effect of road speed and the vacuum of the fan rotation. This suspended particulate travels through the radiator and impacts the fan blades causing a sandblast effect on the fan blades, clutch, and related components assemblies. Riveted fan blades can loosen. Plastic fan blades can become weakened and crack. Fan clutches can become inoperative. Fan blades can dislodge while performing inspections; check blade condition before running engine. Inspect for weak or broken engine/transmission mounts, which can lead to fan blade and shroud damage. Check for proper fan blade clearance, alignment, engagement, and engine/transmission mount function while performing preliminary tests. Also shroud clearance and flow restriction may affect fan efficiency.

19. Inspect radiator (including air flow restriction, missing/ corroded fins, leaks, and damage) and mountings.

Radiators, because of their location mostly at the front of the vehicle, are susceptible to damage from airborne objects and are thus prone to damage or restriction (i.e., bent cooling fins, corrosion, or in the event of large objects, punctures causing loss of coolant).

It is important to inspect the cushion mounts at service intervals to prevent vibration and other movement damage of the radiator and related components. The radiator is usually mounted close to the engine with the intercooler, the air conditioner (A/C) condenser, and oil coolers inhibiting airflow to it. Separate these units at service intervals to inspect for a build-up of compacted road debris that can severely compromise the radiator's ability to effectively cool and rinse out with low-pressure water.

20. Pressure test cooling system and radiator cap.

The normal system pressure for a cooling system is in the range of 15 pounds per square inch (psi), depending on the particular design of the vehicle being tested. The purpose of the radiator cap is to allow a higher operating temperature of the engine by raising the boiling point of the coolant (3°F for each 1 psi). Therefore, it is important to test the radiator cap to ensure that it will retain the indicated pressure setting. Using a pressure tester unit, install the cap onto the tester and proceed to apply pressure to the specified pressure release point. If the specified limit cannot be achieved then cap replacement is required. An underperforming cap may cause engine overheating and coolant loss. If the coolant system has a recovery tank then the cap should have two seals. The one on the end of the coil spring is to contain fluid and the rubber seal under the cap lid is to create a vacuum and aid in returning coolant from the overflow container to the engine.

Pressure testing the engine cooling system for leaks is performed in the same manner. Apply pressure by means of the same tester unit. The gauge reading should achieve the same setting as the recommended radiator cap. The gauge reading should hold steady. The gauge should not increase other than a flutter (water pump pressure surge) upon startup. A large increase usually indicates compression gas infiltration into the cooling system, or if the coolant is oil discolored then check for oil entering the coolant.

21. Inspect coolant hoses and clamps for leaks, damage, and proper routing.

All heater and water hoses should be sound and pliable. All clamps are checked for tightness and/or deterioration. A visual inspection of the entire cooling system is always a good practice. When inspecting hoses, remember that hoses deteriorate on the inside before any sign of wear occurs on the outside. Hoses should be checked by looking inside the hose for cracks and splits.

Silicone hoses are easily damaged by over tightening the clamps. They require special clamps that should be torqued carefully. When correctly installed they provide better sealing and longer service life. Some coolant leaks occur only when the system is cold. They are sealed by expansion during operating temperatures. Check for leaks both when cold and at operating temperature.

22. Inspect coolant recovery system.

The radiator cap will maintain the pressure within the cooling system at or below approximately 15 psi. During heavy loads and prolonged periods of idling, this pressure may exceed the rating of the radiator cap and fluid will be vented to an expansion tank or bottle. Once enough fluid has been vented to reduce the pressure within limits, the spring tension of the radiator cap will close the vent line. At normal temperatures with the pressure cap removed small fluctuations in coolant level can be expected. Inspection of this fluid level may indicate deeper problems. A constant overflowing of the fluid from the tube may indicate air pockets, uneven heating, or hot spots in the engine. Repetitious and rhythmic surges in this level usually indicate a leak to the combustion cylinder, such as a blown head gasket.

23. Identify coolant type; check coolant for contamination, supplemental coolant additives (SCAs), and protection level (freeze point).

Coolant freeze and boil protection levels can be tested with a hydrometer or refractometer. Different antifreeze to water ratios are required to raise or lower the protection level depending upon the temperatures the vehicle will encounter. A 50/50 mixture is most common. To be compatible with the different gaskets and metals that are in various engines, current trucks use one of three types of antifreeze that may be identified by a label in the engine compartment: ethylene glycol, propylene glycol, or carboxylate, which is an extended-life antifreeze. Always ensure that the proper tester is used for the type of antifreeze used. Generally, original equipment manufacturers (OEMs) recommend that the coolant SCA level be tested at each oil change interval.

Additionally, whenever there is a substantial loss of coolant and the system has to be replenished, the SCA level should be tested. Each test provided by an OEM is designed to monitor the SCA package required for its product and cannot generally be used for other OEM's product. Also, the test kits usually consist of test strips that must be stored in airtight containers that have expiration dates that should be observed. Coolant test kits permit the technician to test for the appropriate SCA concentration, the pH level, and the total dissolved solids (TDS). The pH level determines the relative acidity or alkalinity of the coolant. Acids may form in engine coolant exposed to combustion gases or in some cases when cooling system metals (ferrous and copper bases) degrade. The pH test is a litmus test in which a test strip is first inserted into a sample of the coolant, then removed and the color of the test strip indexed to a color chart provided with the kit. The optimum pH window is defined by each OEM, but normally falls between 7.5 and 11.0 on the pH scale. Higher acidity (readings below 7.5 on the pH scale) readings in tested coolant are indications of corrosion of ferrous and copper metals, coolant exposure to combustion gases and in some cases, coolant degradation. Higher than normal alkalinity readings indicate aluminum corrosion and possibly that a low silicate antifreeze is being used where a high silicate antifreeze is required.

24. Service coolant filter/conditioner.

Coolant filters have service intervals and manufacturers' literature should be consulted to determine the proper maintenance schedule. This is important on diesel engines as filters become restricted and coolant additives break down due to extended operation. Not complying with this service information will result in premature engine failure. Some coolant filters contain an SCA charge and corrosion inhibitors. These filters should only be replaced as directed. Never over-service coolant filters that contain an SCA charge.

Supplemental coolant additives are a critical ingredient of the coolant mixture. The actual SCA package recommended by an engine manufacturer depends on whether wet or dry cylinder liners are used, the materials used in the cooling system components, and the fluid dynamics (high flow/low flow) within the cooling system. The operator may wish to adjust the SCA package to suit a specific operating environment or set of conditions. Abnormally hard water, for instance, will require a greater anti-scale protection. Depending on the manufacturer, SCA may be added to a cooling system in a number of ways. The practice of installing the SCA in the system coolant filter is less commonly used today as it resulted in generally higher levels of additives than required and an excess of additives can create problems. Most OEMs suggest testing the SCA levels in the coolant followed by adding SCA to adjust to the required values. Never dump unmeasured quantities of SCA into the cooling system at each PMI.

25. Drain and refill cooling system; bleed air from system; recover coolant.

Cooling system maintenance requires periodic draining and refilling of the system. When doing this it is important to know that all coolants are alike and are not always compatible with each other.

Identify the coolant being used and be sure to follow the pertinent change intervals and refill with the correct type. Extended life coolants (ELCs) have a longer service life than ethylene glycol (EG) and propylene glycol (PG) coolants. When refilling the system it is necessary to bleed the system to allow trapped air to escape. This can be done by opening a hose or fitting on the top of the engine before the thermostat and filling the system until liquid coolant appears. This prevents trapped air from creating a hot spot when the engine is first started. Coolant that has been drained should be handled and stored in accordance with federal, state, and local regulations. This could include on-site recycling or removal by a licensed service. There should be a material safety data sheet (MSDS) posted in your work area that gives information regarding proper handling of each coolant used.

26. Inspect water pump for leaks and bearing play.

The water pump is usually mounted on the front of the engine block with a gasket between the block and the water pump. A leaking water pump seal causes coolant to drip from the drain hole on the bottom side of the pump. Coolant leaking past the water pump seal can lead to bearing failure. A defective water pump bearing may cause a growling noise with the engine idling.

27. Change engine oil and filters; visually check oil for coolant or fuel contamination; inspect and clean magnetic drain plugs.

After the initial lube change at 1,000 to 3,000 miles, subsequent lube changes should be made at the engine manufacturer's suggested intervals for the oil type used, dependent on Class A or AA highway operation. Class A operation is defined as driving on well-maintained highways of concrete or asphalt construction.

Oil filters trap particles that would otherwise cause damage to critical components like engine bearings. Regular replacement of filter elements and inspection of the housing and other component parts is necessary. Modern engines use a variety of materials from cast iron to die cast aluminum for the housing and steel or plastic for machined parts. For this reason, the technician is limited to a visual inspection of filter assembly parts. The visual inspection must be complete and thorough. If there is any doubt, or a component is questionable, replace it. Cracks anywhere in the housing and nicks on machined surfaces can cause major engine failures and, as such, must be detected and appropriate action taken to correct the problem. Some filter assemblies incorporate a bypass valve in their housing. These valve assemblies must be removed and inspected for corrosion, wear, and other signs of damage during overhaul or any time the filter assembly contains metal particles. Canister filters should be cut open and the element material examined for content. Suspended metal particles indicate a potential for bearing failure.

A milky gray sludge indicates water in the oil caused by a possible head or cylinder block leak, which may already have caused bearing damage. Ensure all gasket surfaces are straight and free of nicks, which could cause an improper seal of the assembly. On element-type filter assemblies, ensure that housing grooves, center bolt threads, springs, metal gaskets, and spacers are in good condition. Inspect for distortions in the canister, cracks around the canister bolt hole and clogged passageways in the bolt. Reinstall the bypass valve according to manufacturer's installation procedures and set for proper pressure. Replace the element and all gaskets and supplied seals. When reassembling filter assemblies, apply a small amount of oil to all seals to ensure a tight seal. Pre-filling of canister-type filters is recommended by most OEMs to ensure adequate oil is present during startup to lubricate the bearings.

Used oil and filters are considered hazardous waste and must be disposed of in accordance with federal, state, and local regulations. Used oil filters should never be placed in with other trash but should be disposed of properly. Filter-crushing equipment is available to reduce the volume of disposed of filters. In some states, used oil can be burned in a waste oil furnace, but the appropriate method of disposing is removal by licensed recycling services. Used oil can be harmful to your skin and you should take steps to minimize contact with it. Consult the MSDS sheet posted in your work area concerning the handling of motor oils.

28. Take an engine oil sample.

During a normal PMI, an engine oil sample is taken for inspection of contamination, metal, and dirt content. Also, corrosion-causing agents such as sulfuric and hydrochloric acids and some esters are sometimes tested for content as well. It is for this reason, that the engine oil sample be taken at normal operating temperature to ensure that all possible contaminants and corrosives are measured in the oil in its normal operating condition.

The oil sample should be taken during an oil change at the midpoint of the drainage. The test bottle should be clean before the sample, and sealed and labeled afterward. The label should contain all the pertinent data about the engine and the oil used. Low-level testing can be performed in the shop with a blotter test kit. This kit will compare the darkness and fluidity of a drop of oil on the blotter paper to a code chart that identifies the level of dirt, soot, and particles found in the oil, and the color change of the oil will indicate the acidity. A more comprehensive test can be performed at a specialized lab or oil manufacturer's laboratory. These labs perform spectrographic tests, which identify and quantify the contaminants found in the oil.

B. Cab and Hood (5 Questions)

1. Inspect key condition and check operation of ignition switch.

Inspect the vehicle keys for wear, nicks, bends, or other damage. Ensure that the ignition switch is securely mounted and operates smoothly without binding and provides constant power at the accessory, on, and crank positions. If the truck is equipped with a push-button start switch, it should operate easily without sticking while properly closing the cranking circuit when depressed. Some trucks with keyless ignition systems require a driver ID code or card instead of a key.

2. Check operation of indicator lights, warning lights, and/or alarms.

Warning lights have changed significantly over the years. Earlier lamps were small, round lights that were color coded for their system; yellow for a warning or caution in a non-safety-related system and red for safety-related system warnings. Major or safety-related systems may still use lights but many non-safety systems use light-emitting diodes (LEDs) behind a pictogram of the system it is monitoring. This improvement allows for less electrical current usage in the instrument panel and more reliable operation. Most vehicles have a self-check feature that illuminates all warning lamps when the ignition switch is first switched to the on position. This lets the driver and/or technician know that the lamp and circuit is operational. Alarms or buzzers are also used for some systems such as low air pressure (which is mandatory) and for low oil pressure. The low air pressure alarm must sound when the system air pressure drops to below 50 percent of the governor cut-out pressure (approximately 60 psi).

3. Check operation of instruments/gauges and panel lighting.

A short circuit in the sending unit wire to the gauge will lower resistance in that circuit, causing the gauge needle to read high (see the following illustration). Generally, a short to ground in a sending unit wire will cause an overload of current to the gauge, which will consequently burn out in a very short time.

Although not considered official criterion for out-of-service (OOS) status, a malfunctioning oil pressure gauge may be cause for an inspector to deadline a vehicle because a no-oil pressure condition can rapidly destroy an engine. Also, a proper and thorough check of a vehicle's instrument panel is always an important part of any PMI. When the ignition is first turned on, electronic gauges perform a self-test by sweeping the needles.

4. Check operation of electronic power take off (PTO) and engine idle speed controls (if applicable).

Modern diesel engines manage idle speed and shutdown by computer control. This is sometimes a PTO function when the engine speed is set to operate auxiliary equipment. Idle speed usually can be raised and lowered by the use of a toggle switch provided on the dash. During an inspection, check the operation of this switch to ensure that idle speed can be controlled. If equipped with a PTO switch, also check its operation.

Engine shutdown can be programmed to occur after a desired time interval. When the desired time has elapsed without input from the throttle pedal or the clutch pedal the engine will shut down automatically. To change this programmed time, an electronic service tool (EST) is needed. Most engine management systems have automatic stop/restart features that expand the normal idle shutdown feature to further reduce idle time. They will automatically stop and restart the engine when required to hold engine temperature above a programmed specific value, maintain the battery charge level, and to keep the cab/sleeper at a temperature level programmed into customer parameters. Other benefits claimed are reduced emissions, noise, and engine life. In order for the stop-restart feature to be enabled, a number of conditions must be met. One OEM system requires that the ignition switch be on, the engine idling with the idle shutdown timer enabled, the hood closed, the transmission in neutral or park, the parking brake set, and the master cruise switch in the on position. When the system determines to start the engine, based on sensor inputs, the active light illuminates and an alarm in the engine compartment sounds and the engine cranks. In the event the engine fails to start, a second attempt will occur in 45 seconds. At startup, engine RPM ramps up to 1,000 rpms. If the restart was prompted by input from the cab temperature sensors, after about 30 seconds, the climate control system will turn on to manage cab/sleeper temperature. Engine shutdown can also occur to prevent damage to the engine from lack of lubrication or overheating conditions. The way in which the engine electronics respond to a potentially damaging condition is known as failure strategy. Failure strategy response levels are customer programmable and typically would be driver-alert (lowest level), engine ramp-down, and engine shutdown (highest level). Manufacturers recommend programming the highest level failure strategy response but, for obvious reasons, emergency vehicles such as fire trucks are usually programmed with low-level failure strategy. An EST can be used to check the programmed failure response.

5. Check operation of defroster, heater, ventilation, and A/C (HVAC) controls.

Check the HVAC system to see if it is able to regulate temperature in the cab and sleeper. The blower motor should operate in all speeds. The blower should produce an adequate flow of air and should not be abnormally noisy. Move the mode selector, making sure that air is directed to the selected outlets. A pocket thermometer can be used to check the outlet temperature of the heater and the defroster. With the engine idling, the technician should hear the A/C compressor clutch engage when air conditioning is selected. Raise the engine to about 1,500 rpm and close the doors and windows. Turn the blower to high and select the coldest temperature. Run the system for about 10 to 15 minutes and compare the temperature of the air at the outlet nearest the evaporator with specifications in the service manual. Check the operation of the sleeper controls in a similar manner. If the truck is equipped with automatic temperature control (ATC), the outlet temperature should be within several degrees of the set temperature.

6. Check operation of all accessories.

Identify all the cab and chassis accessories. Where possible, switch on to verify operation. Many accessories can be checked during a road test. For instance, a vehicle on-board radar (VORAD) collision warning system (CWS) has to be tested on the road as well as with an EST.

7. Using diagnostic tool or on-board diagnostic system to read current and historic diagnostic trouble codes from electronic modules (including engine, transmission, brake, supplemental restraint, traction control and body control systems).

The typical process for checking DTCs is to use a handheld EST or laptop computer to access the databus; then check for, record, and clear DTCs. These systems are controlled and monitored by electronic modules. Most systems will accommodate a snapshot readout from the electronic control module (ECM) to facilitate troubleshooting intermittent problems that either do not generate codes or do not store a code without a clear reason.

Control modules store fault codes against parameter identifications (PIDs). To access the databus, the technician connects a diagnostic reader to the American Trucking Association (ATA) connector. Once the diagnostic reader is connected, most systems allow two-way communications for reading system parameters, calibration data, diagnostics, data programming, recalling faults, and audit trails. Some electronic systems can display active fault codes by blinking two-digit numerals on the instrument panel indicator. Diagnostic tests are menu-driven and should be used to identify the FMIs (Fault Mode Indicators). To test a system, once the problems have been corrected, the now inactive fault codes need to be cleared. ESTs can be used to clear these codes.

ESTs capable of reading the chassis databus are connected to the on-board electronics by means of an SAE/ATA J1587/J1708/1939 6- or 9-pin Deutsch connector in all current systems. This common connector and the adherence by the engine electronics OEMs to SAE databus software protocols enables proprietary software of one manufacturer to at least read the parameters and conditions of their competitors. This means that if a Caterpillar-powered truck has an electronic failure in a location where the only service dealer is Detroit Diesel Corporation (DDC), some basic problem diagnosis can be undertaken using the DDC electronic diagnostic equipment.

There are five categories of ESTs as follows:

1. On-board diagnostic lights blink or flash codes (OEMs use both terms) are an on-board means of troubleshooting using a dash or ECM-mounted electronic malfunction light or check engine light (CEL). Normally only active codes (ones indicating a malfunction at the time of reading) can be read in truck (engine) ECMs but some will also read historic codes.

2. Scanners are read-only tools capable of reading active and historic or inactive (logged but not currently indicating a malfunction) codes and sometimes system parameters, but little else. They are obsolete as a truck/bus diesel engine diagnostic tool.

3. Generic reader/programmers are microcomputer-based ESTs designed to read and reprogram all proprietary systems in conjunction with the appropriate software cartridge or optical card. They are usually tough and portable. The ProLink 9000 has become the industry standard.

4. Proprietary ESTs are usually PC-based test instruments packaged by the OEM for use exclusively on their system. They have the advantage of offering an optimum degree of user friendliness and the disadvantage of being system specific and high in cost. Examples are:

   ■ Cummins: Compulink, Echeck, INSITE™
   ■ Cat: Electronic Technician (ET)
   ■ DDC: Detroit Diesel Diagnostic Software

These proprietary ESTs have fallen out of favor due to the near universal acceptance of generic PCs.

Most OEMs have made the generic PC and its operating systems (Microsoft Windows) their diagnostic and programming tool of choice. PCs are easily upgradable and have vast computing power when compared with proprietary reader/programmers. OEMs can upgrade/update software more easily and the PC is not confined to the limitations of all the other existing electronic troubleshooting tools. PCs are connected to vehicle ECM through the SAE/ATA J1708/1939 6- or 9-pin connectors; a serial link or interface module may also be required, depending on the age of the system.

Generally, SAE J1587 covers common software databus protocols in electronic systems, SAE J1708 covers common hardware protocols, and the more recent SAE J1939 covers both hardware and software protocols. The acceptance and widespread usage of these protocols enables the interfacing of electronic systems manufactured by different OEMs on truck and bus chassis, plus permits any manufacturer's software to at least read other OEMs' electronic systems.

8. Check operation of electric and air horns.

Electric and air horns are important safety devices enabling the operator to warn other motorists of imminent danger. Both are required to be fully functional. They should be tested during the driver's pre-trip and post-trip inspections. Verify their operations as part of each preventive maintenance inspection.

9. Check condition of safety equipment, including flares, spare fuses, reflective triangles, fire extinguisher, and all required decals.

Check that the truck is equipped with all the necessary safety equipment. The truck should have spare fuses (unless equipped with circuit breakers), three red reflective triangles, and a properly charged and rated fire extinguisher within arm's reach of the driver's seat. Flares, lanterns, and flags are optional. CVSA out-of-service criteria calls for deadlining when safety devices such as chains and hooks have been repaired improperly. Federal regulations prohibit the usage of a vehicle that does not have at least one light illuminated for loads projecting more than 4 feet beyond the vehicle. Rear chains and hooks that are used in double trailer systems for hookup should not be worn to the extent of a measurable reduction in link cross sectional area or have any significant abrasions, cracks, or other faults that would affect structural integrity. If any defects are found, deadline the vehicle.

10. Inspect seat belts, seat tethers, and sleeper restraints

Check the operation of all seat belts and restraints. With the seat belt on and the truck moving, activate the parking brake to brake the truck very quickly and determine if the seat belt inertia reel locks. Visually inspect seat belt anchors, seat bolt downs, and sleeper bunk harnesses.

11. Check supplemental restraint system (SRS) for proper warning light operation and diagnostic trouble codes (if applicable).

If the vehicle is equipped with the supplemental restraint system (SRS), you should also check the operation of that. This is done easily because whenever the key is turned on, the SRS performs a self-test. You should observe that the "SRS" or "Airbag" indicator light illuminates and stays on for several seconds following startup—then extinguishes. If it stays on continuously or blinks, there is a problem in the system that needs correction.

12. Inspect wiper blades and arms.

Check the wiper blades and arms for correct operation. The blade should remove water from the entire sweep area. Replace the blade if the rubber is cracked, separated, has segments missing, or is stiffened with age and cannot properly clear the windshield. The arm attaches securely to the wiper mechanism and provides the correct tension for the wiper blade to windshield contact. The arm should be replaced if it cannot perform as stated above or if it is bent.

13. Check wiper and washer operation.

Check the wiper and washer systems for proper operation. Ensure that the wiper blades and arms are in good condition and that the linkage is lubricated and in good working order. Check electric wiper motor operation through all the speeds and delay settings. If any speeds are inoperable, follow the manufacturer's service procedures to identify the failed component and its replacement procedures. Air-operated wiper motors should be checked as previously mentioned but with additional checks for air leaks. Adjust the arm to sweep and park in the correct areas and not contact the cab in any position. Inspect the washer fluid container for debris that could impair the future operation of the system and clean if necessary. Ensure that the correct washer fluid is in use for the season and conditions that the vehicle is operating under. Check the condition and aim of the spray nozzles and perform any necessary adjustments or required replacement of defective components. CVSA safety standards out-of-service criteria state that any wiper unit that has inoperable parts or missing wipers that render the system ineffective on the driver's side shall be cause for deadline.

14. Check for all required vehicle permits, registration, decals, and inspection papers.

Every vehicle must have a registration (license) in order to operate. The majority of trucks or truck-trailer combinations weighing more than 26,000 pounds will be registered under the International Registration Plan (IRP). A cab card is issued to the vehicle, stating the IRP areas in which the vehicle is allowed to operate. The Federal Highway Administration has set up a minimum inspection standards program, under which each vehicle must carry proof that an inspection was completed. Proof can either be a copy of the inspection form kept on the vehicle or a decal. If using a decal, a copy of the inspection form must be kept on file and must indicate where an inspector can get a copy of it. The only requirement of a decal itself is that it be legible. Each vehicle must be inspected separately. This means a tractor/trailer is considered to be two vehicles, each requiring a decal; a converter dolly is a separate vehicle. The decal must show the following information: date vehicle passed inspection, name and address to contact concerning inspection records, and the VIN.

15. Inspect windshield glass for cracks, chips, clarity, discoloration/glazing, or other damage; check sun visor operation.

Check the windshield glass for cracks, dirt, illegal stickers, or discoloration that obstruct the driver's view. Clean and adjust as necessary. Any crack over ¼-inch wide, intersecting cracks, discoloration, or other vision-distorting matter in the sweep of the wiper on the driver's side will take a vehicle out of service by the commercial CVSA out-of-service standards.

16. Check seat condition, operation, mounting, and suspension components.

Many cab air seats contain one or two air springs, two shock absorbers, and a leveling valve. The leveling valve maintains the proper air pressure in the air springs to provide the correct seat height. An air-suspended seat reduces road shock transferred through the chassis, cab, and seat to the driver. Air pressure is supplied to the air-suspended seat through a pressure protection valve and a regulator.

17. Check door glass and window operation.

Check to see if the windows will go up and down. Windows must go up and down when actuated electrically or by hand.

18. Inspect steps and grab handles.

Inspect that all steps are in good condition and secure. Check that the grab handles are also secure and tight to the cab.

19. Inspect mirror mountings, brackets, glass, heaters, and motors.

A typical fault such as a loose mirror mounting is always recorded on a PMI. There should be no blemishes on the mirror glass. This is in accordance with the Department of Transportation's Federal Motor Carrier Safety Regulations and the CVSA standard inspection procedures. If the truck is equipped with either heated or motorized mirrors, make sure they are functioning. When the mirror heat switch is on, an indicator light should be on and the mirror surface should become warmer than room temperature.

20. Inspect and record all observed physical damage.

Check the vehicle area for hazards on the vehicle and any physical damage such as dents or cracks. Performing a driver's circle check is a good way of doing this. Use a combination of visual and hands-on inspection. The inspection should start at the left front of the vehicle and work in a counterclockwise direction around it.

21. Lubricate all cab and hood grease fittings.

Use light machine oil and a lithium-based grease on hinges and latches.

22. Inspect and lubricate door and hood hinges, latches, strikers, lock cylinders, linkages, and cables.

Lubrication of door latches, locks, and hinges are always performed as part of a PMI. Light machine oils and lithium base grease can be used.

23. Inspect cab mountings, hinges, latches, linkages; service as needed.

Most current cabs use air ride. Check the foreword pivots for wear and excessive movement, check the ride height and leveling valve, inspect shock absorbers for leakage. Visually and dynamically test the bushings.

24. Inspect tilt cab hydraulic pump, lines, and cylinders for leakage; inspect tilt cab safety devices; service as needed.

With most hydraulic cab lift systems there are two circuits: the push circuit that raises the cab from the lowered position to the desired tilt position and the pull circuit that brings the cab from a fully tilted position up and over the center.

Inspect the hydraulic cylinders for external leakage and determine if the cab can be raised and lowered. Check for internal leakage by observing bleed down. Make sure the mechanical stops engaged and hold safely.

25. Check accelerator, clutch, and brake pedal operation and condition.

Check the accelerator linkage for binding and lubricate it at the appropriate locations. Test the brake pedal for operation and lubrication. Check clutch-pedal freeplay and if an adjustment is performed the work must be noted using a signoff or work performed sheet before the vehicle can be put back into service. If the problem noted did not require repair, but was still looked at, then this must be noted as well. A common problem in electronic accelerator assemblies is dirt build-up around the throttle position sensor (TPS) assembly. Electronic fuel-injected diesel and gasoline engines use TPSs. The TPS is a variable resistor that supplies the ECM with throttle position information so the ECM can regulate engine performance and is located in the electronic foot pedal assembly (EFPA). Many systems use a 5-volt reference voltage that is sent from the ECM to the TPS, and the TPS returns a portion of it, proportional to pedal mechanical travel. The actual voltage signal returned to the ECM is converted to counts, which can be read with an EST.

26. Check cab ride height; inspect cab air suspension springs, mounts, hoses, valves, shock absorbers, and fittings for leaks and damage.

When the truck is on a level surface the cab should also appear level. To check the cab air suspension, make sure that the air brake system pressure gauge reads the specified system cut-out pressure. Use a tape measure to check cab ride height at the location specified by the manufacturer. This is usually between the upper and lower plates. Compare your readings with specifications and adjust the leveling valve as necessary. Use a soap and water solution to check for leaks. Also check the condition of the shock absorbers in the cab suspension system by looking for fluid leaks and observing any roughness in the ride when the cab is at normal ride height. Also, check the condition of the transverse rod and mounts by looking for excess lateral movement.

27. Inspect front bumper, fairings, and mounts.

Check the front bumper to make sure that it is securely mounted and is not cracked or otherwise damaged. Make sure that there is sufficient space between the ends of the bumper and the front tires to avoid damage to the tires. The fairings should also be checked for cracks and that the mounts are tight and undamaged.

28. Inspect A/C condenser and lines for condition and visible leaks; check mountings.

The technician/inspector must report and log all refrigerant leaks into a PMI schedule, so these repairs are addressed in a timely manner. All technicians that perform service on an A/C system of a motor vehicle must be certified, regardless of the capacity of the person performing the service (i.e., shop foreman, parts specialist, and technician).

The condenser is normally mounted just in front of the truck radiator, or if rooftop-mounted, in the center of the roof above the driver/passenger area. In either position, it receives the full flow of ram air from the movement of the truck.

Refrigerants carry oil with them and when they leak, there is usually an oily film around the source of the leak. Observe hoses, fittings, and seals for signs of leakage. Sometimes a dye has been added to the system to help locate leaks. These dyes are designed to be seen using a black-light leak detector. Smaller leaks may require the use of electronic leak detectors. Check the condensers mounting bolts and rubber insulators. A condenser that is allowed to move or is subject to vibration may lead to more serious leaks.

29. Inspect A/C compressor and lines for condition and visible leaks; check clutch; check mountings.

Depending upon the rate of leakage, the Clean Air Act establishes regulations concerning the repair of leaks. The technician/inspector must report and log all refrigerant leaks into a PMI schedule so these repairs are addressed in a timely manner (see the following illustration).

30. Check A/C system condition and operation.

Begin checking the A/C system condition and operation with a preliminary inspection. Check the compressor drive belt condition and tension. While checking the belt, look at the compressor mountings, checking for loose bolts or cracked mounts. Use a thickness gauge to check the compressor clutch condition. Visually check for damaged or leaking hoses and road debris buildup on the condenser coil fins. If the condenser fins are becoming plugged with debris, use compressed air, soapy water, a small brush, and a fin comb to clean the fins and restore airflow.

Engage the compressor with the engine idling and listen for indications of mechanical, pressure, or charge level problems. Some R-12 systems are equipped with a sight glass that may be used to check system charge, but R-134a systems usually do not. The use of a sight glass to check level of charge is not always conclusive and is being replaced by temperature testing by some technicians.

Follow the OEM-recommended procedure for troubleshooting the system. Frost on the receiver/dryer usually indicates an internal restriction in that component. Because the receiver/dryer is located in the high-pressure liquid line between the condenser and evaporator, it should feel warm. Thermal expansion valve (TXV) frosting indicates a restricted valve or one sticking closed. Frost formation on the evaporator outlet indicates a flooded evaporator, caused by an excessive refrigerant charge, or a stuck-open TXV. Unusual noises can often guide the technician to a faulty component. These problems may also cause frost on the compressor suction line.

On systems using an accumulator, it should feel cold because of its close proximity to the evaporator. Both the evaporator inlet and outlet should feel cold when operating normally. On orifice tube systems, the evaporator inlet should be slightly warmer to the touch than the outlet. If the evaporator outlet is warm, the refrigerant charge may be low. High-side refrigerant components should feel hot or warm, and low-side components should feel cool or cold. A plugged drain in the evaporator case may cause a sulfurous odor in the cab. R-12 has no odor, while R-134a has a faint ether-like smell.

31. Check HVAC air inlet filters and ducts; service as needed.

Change HVAC air inlet filters at least once a year, preferably at the beginning of summer. A clogged evaporator core drain will cause windshield fogging or a noticeable mist from the panel vents. A clogged drain normally can be opened with a slender piece of wire or with low-pressure shop air. The evaporator drain should be checked during routine maintenance inspections. A cracked evaporator case can cause a whistling noise during high blower operation. Minor cracks can be repaired using epoxy-type adhesives. A mildew smell that is noticeable during A/C system operation can be rectified by removing the evaporator case and washing it with a vinegar and water solution or a commercially available cleaner.

32. Check auxiliary power unit (APU) operation and mounting; inspect for damage, leaks, and belt tension.

APUs provide electrical power, cooling, and/or heating for the truck without operating the truck's engine. Routine maintenance depends on the type of APU used. Engine-based systems need oil, filters, and possibly drive belts at recommended maintenance intervals. The fuel system, coolant hoses/clamps, belt tension, exhaust, mounting brackets, and so on need to be performed at regular intervals as well.

C. Electrical/Electronics (10 Questions)

1. Inspect battery box(es), cover(s), and mountings.

An inspection of a battery should always begin with a thorough visual inspection of the case and the immediate surrounding area for any signs of leakage. Check mounting bolts, hold-down clips, and cover clamps. Rubber hold-downs and insulators should have no tears or cracks.

2. Inspect battery hold downs, connections, cables, and cable routing; service as needed.

Battery terminals are either two tapered posts or threaded studs on top of the case or two internally threaded connectors on the side. Some newer batteries have both types of terminals so that one battery fits either application. When disconnecting a battery, the ground cable should always be disconnected first, then the positive cable; therefore, when reconnecting a battery, the positive cable should always be connected first and the ground cable last. Heavy-duty trucks typically use three or four batteries connected in series or parallel to allow operation of a 12-volt or 24-volt electrical system, and/or to increase the cold cranking amperes (CCA) rating of the vehicle. The most common hookup is the parallel hookup, which can be further described as a single-path circuit or dual-path circuit. The path is often determined by the placement of batteries, such as two batteries per side of truck or all the batteries in one battery box on one side of the truck or behind the cab.

Check the condition of the battery cables by performing voltage drop tests. Connect the voltmeter in parallel with the portion of the circuit being tested, being careful to observe polarity. For example, to test for resistance caused by battery terminal corrosion, place one voltmeter lead on the battery post, the other one on the cable, and then observe the meter while cranking the starter. A good connection at the battery post will show that only about 0.1 volt is dropped. When the voltage dropped across the entire positive battery cable is measured there should be less than 0.5 volts dropped. You can test the positive battery cable by placing one lead on the battery post and the other on the solenoid battery cable stud, testing voltage drop through the entire cable. A similar test can be used to test the ground circuit: place one lead on the negative battery post and the other on the starter frame and observe the voltage reading while cranking. There should be less than 0.3 volt dropped across the negative cable. All wires somewhat resist current flow. The value of this resistance is relative to the size and type of wire used. In relation to current flowing through the wire, the smaller the wire is, the greater the resistance value. Resistance generates heat, which further increases resistance. A battery can be cleaned with a baking soda and water solution. Always wear hand and eye protection when servicing batteries and their components. If the built-in hydrometer indicates light yellow or clear, the electrolyte level is low and the battery should be replaced. A high-voltage regulator setting that causes overcharging may cause the low electrolyte level. It is a good practice to spray the cable clamps with a protective coating to prevent corrosion. A little grease or petroleum jelly will also prevent corrosion. Also available are protective pads that go under the clamp and around the terminal to prevent corrosion.

3. Check and record battery state of charge (open circuit voltage) and condition.

The first step is a visual inspection of the battery case, terminals, and electrolyte level before testing can be performed. If there is any leakage from the case due to cracks or loose posts/terminals then the battery is deemed unserviceable and must be replaced.

After this the level of the electrolyte (a combination of sulfuric acid and distilled water) must be serviced if required. The plate area must be completely immersed in electrolyte. If the level is low (indicated by the plate area being exposed), then distilled water only can be added to return the electrolyte to proper balance. Tap water contains impurities and will lead to premature battery failure.

A battery having moisture on the top surface, a strong odor of sulfur, expanded case or excessively high temperature usually indicates overcharging. Any moisture on the battery surface will cause a discharge condition and must be removed. Cold-weather starts are critical to the operation of a truck. A battery that is even partially discharged will be exposed to a freeze-up problem in cold weather. The specific gravity in a discharged battery is decreased to a point where the electrolyte can freeze.

A battery open-circuit test can be used to determine its state of charge. A battery that has recently been charged by running the engine, or using a battery charger, must have its surface charge removed before testing. Surface charge may be removed by applying a load of 300 amps for 15 seconds with a carbon pile load tester, or by cranking the engine for 15 seconds. After removing the surface charge, stabilize the voltage by allowing the battery to rest for 15 minutes. The following chart may be used to correlate voltage with state of charge.

State of Charge Determined by Open Circuit Voltage Test	
Stabilized Open Circuit Voltage	**State of Charge**
12.6 volts or more	Fully Charged
12.4 volts	75% Charged
12.2 volts	50% Charged
12.0 volts	25% Charged
11.7 volts or less	Discharged

4. Perform battery test (load and capacitance).

If the battery passes the state of charge test (3) then perform a load test to determine if the battery is capable of providing the proper discharge amperage and voltage under load conditions.

> *Note:* A two-minute period should be maintained between any load testing cycles to avoid damage to the battery.

If this is a multiple-battery vehicle, then first disconnect the batteries and perform individual testing of each battery. Determine the amperage rating of the vehicle (check manufacturer's specifications) and confirm that the proper batteries are installed in the vehicle. If the voltage is below 12.4 volts DC at 70°F, the battery should be recharged to above 12.6 volts. Apply a load through the tester at one-half the CCA for 15 seconds. If the voltage drops below 9.6 volts at 70°F, the battery is unserviceable.

> *Note:* A quick test can be performed if the battery voltage indicates that it has to be recharged.

As charging to proper levels can take hours to complete properly, it would be good time management to install the battery charger, set it to a medium charge rate (voltage), amperage approximately 40 amps, and observe the voltage rate. If the voltage exceeds 15 volts DC within two to three minutes then the battery requires replacement.

> *Note:* For safety and accuracy, testing cannot be performed on frozen batteries. Also, note that the sulphated battery test just mentioned may not be accurate on all battery types.

Voltage/Temperature Table

Minimum Voltage	Temperature	
	Degrees Fahrenheit	Degrees Celsius
9.6	70	21
9.5	60	16
9.4	50	10
9.3	40	4
9.1	30	−1
8.9	20	−7
8.7	10	−12
8.5	0	−18

5. Inspect starter, mounting, connections, cables, and cable routing.

Starter motors produce high torque, so check the clamping torque of the starter flange fasteners. Components of the starting circuit are the battery cables, magnetic switch or solenoid, and the starter motor. Starter relays may be integral or non-integral with the starter. External starter relays (also known as a "mag" switch) can attach either to the starter as a separate unit or to the firewall. Removal of these types of starter relays entails removing the two positive leads and the control wires. The starter relay is an electrical high-current switch and, as such, it does not have a ground lead. If the starter system has a magnetic switch, use a jumper wire around the heavy terminals. If it cranks, the relay is defective assuming that control current from the starting switch is available at the small terminal of the relay. Any time maintenance is to be performed on the starter, remove the ground (negative lead) from the battery to prevent accidental shorting of the wiring. A voltmeter is connected across a component in a circuit to measure the voltage drop across the component. A current must be flowing through the circuit during the voltage drop test. If a circuit has a normal resistance, the voltage drop is within specifications. For example, with the engine cranking, the positive battery cable may have a 0.2-volt drop. Excessive resistance in a cable or component causes a higher than specified voltage drop and reduced current flow. The sum of the voltage drops in any series circuit will equal the source voltage or applied voltage.

6. Engage starter; check for unusual noises, starter drag, and starting difficulty.

While cranking the engine, listen for unusual sounds that could be indications of problems. If the starter is noisy while cranking the engine there may be loose or broken mounting bolts, or the teeth on the flywheel or starter drive may be worn or damaged. The starter may spin without cranking the motor, indicating a worn starter drive or damaged teeth. An engine that cranks slowly or unevenly can result in a no-start condition because a diesel engine has to reach a minimum cranking speed before it can start.

Whenever the engine cranks over too slowly remember that it could be caused by engine mechanical problems or it could be related to the starter. The starter may be receiving insufficient voltage or current from the battery. Other noises to listen for are a starter that chatters and disengages before the engine has started and a starter that engages but does not spin. Note when any of these noises are present, so that more thorough testing can be performed to find the cause of the problem.

High starter current draw, low cranking speed, and low cranking voltage usually indicate a faulty starter motor. Low current draw, low cranking speed, and high cranking voltage usually indicate excessive resistance in the starting circuit. The starting system circuit carries the high current low within the system and supplies power for the actual engine cranking. The cranking current test or starter draw test measures the amount of current, in amperes, that the starter circuit draws to crank the engine. This amperage reading is useful in isolating the source of certain types of starter problems. An inoperative starter, along with a high current draw is always an indication of a direct short to ground. Any opens, failed switches, springs, or relays would cause an operative failure and a no-current condition.

7. Inspect alternator, mounting, wiring, and wiring routing.

If the alternator mounting holes are distorted due to over-torquing and/or age, then the alternator should be replaced. Severely out-of-round mounting holes on an alternator will cause misalignment of the drive belt and may cause premature bearing failure. A complaint of an unusual noise requires a thorough visual inspection. A visual inspection will usually determine whether a loose component is causing a vibration or any unusual sounds. Drive belt tension also is critical to prevent slippage.

8. Perform alternator current output test.

The minimum output requirement of the alternator (voltage and amperage) must first be confirmed to the manufacturer's specification. Determine if requirements will exceed output (i.e., due to additional lighting, trailer, refrigeration unit). Note: All testing is usually performed at 5,000 alternator rpm or 1,500 to 2,000 engine rpm for purposes of testing accuracy. If the vehicle is a late model and computer controlled, this speed may not be possible as the engine ECM will not permit anything above a preset limit when there is no load on the engine. Perform battery testing as outlined in 4 prior to this test. Using a load tester unit, place the positive cable of the tester onto the positive battery cable and the negative cables onto the negative battery cable. Then place the inductive pickup clamp of the tester around the output cable of the alternator. The load tester should now indicate battery voltage. Start the engine, and raise engine speed to the specified rpm. Apply a load using the carbon pile until the specified maximum amperage is read without dropping system voltage below 13 volts. Alternator output should be within 10 percent of the specified rating. As ambient temperature increases, alternator output may decrease.

9. Perform alternator voltage output test.

This voltage test can be used in conjunction with 8 as the amperage output of the alternator will have an effect upon the voltage being produced. Information should be confirmed as to the manufacturer's specifications/settings. Voltage output of the alternator is to be measured (engine running) at the alternator output terminal using a voltmeter. Output will depend on load. The in-dash type voltmeter is usually connected at or near the battery. To verify voltage accuracy of the dash voltmeter, connect the test meter on the battery terminals in proper polarity; the meter reading of the test and dash unit should be within a ½-volt DC of the test unit.

10. Check operation of interior lights; service as needed.

Check the operation of the interior lighting. Although inoperative interior lighting is not considered a basis for placing a vehicle out of service, it is still important for safe operation of the truck. The dome lights and luggage compartment lights are usually operated by a door jamb switch that completes the ground circuit. When the door is opened, the light should come on. Some interior compartment lights use a mercury switch that closes when the door is opened. Dash lighting should operate when the lights are on and the dimmer should adjust them to the desired level. Common lighting problems are power supply, ground, and bulbs. Making sure that the bulb is good, that there is adequate power, and that there is a good ground path will solve most problems. When more than one light has similar symptoms, such as not operating or being dim, look for what they have in common. For example, if all of the interior lights are dim, the cab may not be adequately grounded. This is especially true when lights are attached to plastic panels, and the ground must be completed elsewhere.

Continuity in an electric circuit may be tested with a high-impedance digital voltmeter. Connect the meter ground lead to a satisfactory ground connection and connect the positive meter lead to various locations in the circuit starting at the voltage source. When the meter reads a very low voltage, the open circuit is between the meter positive lead and the previous location where a normal voltage reading was obtained. A 12-volt test light may be used to test continuity in a circuit, but a high-impedance test light must be used when testing continuity in computer-controlled circuits. If an ohmmeter is used to test continuity, the meter leads must be connected across the terminals on the component to be tested with the component disconnected from the circuit. An ohmmeter may be damaged if it is connected to a circuit with voltage supplied to the circuit. Always disconnect the truck's battery before using a self-powered test lamp to test continuity. Never use a self-powered test lamp on an electronic module circuit, as damage to the module may result.

11. Check all exterior lights, lenses, and reflectors; check headlight alignment; service as needed.

Exterior lights are generally mounted on the sides, roof, and front of the vehicle. On the sides (depending on cab design) are turn-signal lights, side-marker indicators, and sometimes an intermediate turn signal. On the roof are clearance lights (usually on either end), identification lights (usually in the middle), and sometimes additional utility lights. In the front are single or dual headlights, fog lights, turn-signal lights, and sometimes additional utility lights.

In order to properly adjust the headlights, headlight aim must be checked first. Various types of headlight-aiming equipment are available. When using aiming equipment, follow the instructions provided by the equipment manufacturer. Where headlight-aiming equipment is not available, headlight aiming can be checked by projecting the upper beam of each light upon a screen or chart at a distance of 25 feet ahead of the headlights. The vehicle should be exactly perpendicular to the chart.

An operational check of the brakelight switch is performed during a check of the electrical system, since the brakelight or stoplight switch, as it is generally called, is part of the vehicle's electrical system. Defective and/or improper ground connections are the most common sources of dim lights. In addition to replacing all defective lamps and bulbs, periodically check to see that all wiring connections are clean and tight, that lamp units are tightly mounted to provide a good ground and headlights are properly adjusted.

12. Inspect and test trailer power cord connector, cable, and holder; service as needed.

The junction block for the trailer lights is located inside the cab, often directly behind the driver seat. Access is usually gained by removing a plastic cover held in place by four screws. In sleeper models, the junction block is usually in the luggage compartment. Whenever a lighting inspection of a vehicle and/or combination vehicle(s) is to be performed, the inspector usually starts at the front and works back from there. This is because the source of power is usually up front. In most modern box trailers, there are no splices for corrosion to attack and the trailer junction box has been eliminated. Most current vehicles use a seven-wire trailer plug and connector. Ensure that the prongs and sockets are not corroded.

D. Frame and Chassis (22 Questions)

1. Brakes (10 Questions)

1. Check air parking brake operation.

The parking brake must be capable of holding a fully loaded truck. A truck with a park brake that will not hold is dangerous, and should not be dispatched. Air brake equipped trucks will have park brakes on one or more of the non-steering axles. They are applied by spring pressure and released by air pressure. Start the engine and wait until the air system reaches system pressure. Push in the yellow-colored, diamond-shaped dash control valve knob to release the park brakes. It should stay in; if it doesn't, check for a pressure loss or a defective control valve. Test the park brakes by putting the truck in low gear, moving it forward at idle speed, and applying the park brakes by pulling out on the knob. The truck should stop promptly. If it does not, repairs are needed. Park brake performance is affected by brake free-stroke adjustment. Excessive free stroke will reduce both park and service brake performance.

2. Check and record air governor cut-in and cut-out settings (psi).

The governor manages compressor loaded and unloaded cycles. It defines the system pressure. System pressure in most trucks is set at values between 120 and 135 psi, with 125 psi being most typical. System pressure is any value between cut-in and cut-out pressure. Cut-out pressure is the pressure at which the governor outputs the unloader signal to the compressor. The unloader signal is maintained until pressure in the supply tank drops to the cut-in value. Cut-in pressure is required by Federal Motor Vehicle Safety

Standards (FMVSS) 121 to be no more than 25 psi less than the cut-out value. The difference between the two values' range is between 20 and 25 psi.

Governor operation can be easily checked. One method is to drop the air pressure in the supply tank to below 60 psi and, with the vehicle's engine running, build up the pressure. A master gauge should be used to record the cut-out pressure value. This should be exactly at the specification value. If not, remove the dust boot at the top of the governor, release the lock-nut and turn the adjusting screw either clockwise to lower or counterclockwise to raise the cut-out pressure. Governors seldom fail but they do not have check valves. If the unloader signal is not delivered to the compressor unloader assembly, high system pressures will result. If the safety pop-off valve on the supply tank trips (this occurs at 150 psi), this is usually an indication of governor or compressor unloader malfunction. The only adjustment on an air governor is the cut-out pressure value. If the difference between governor cut-out and cut-in is out of specification, the governor must be replaced.

3. Service air drier as needed. Check air drier purge valve operation and air drier heater, if equipped.

Water and oil residue collected in the air dryer must be purged. Each time the compressor unloader cycles, it triggers the purge valve in the air dryer sump, ejecting moisture and any accumulated sludge. The purge valve may become clogged and need to be cleaned. During cold weather, it may freeze up if its heater fails. The heater should operate when the temperature falls below about 40°F. Test the operation of the purge valve by cycling the compressor through several cut-in/cut-out cycles, and observing the operation of the purge valve. Each time the unloader cycles, there should be a discharge of air and moisture from the bottom of the air dryer.

4. Check air system for leaks (brakes released).

First, block the wheels on the vehicle. Build air pressure to governor cut-out, release the parking brakes and shut off the engine. Walk around the unit listening for audible air leaks. Then check dash gauges for system pressure loss. Investigate any out-of-specification loss. Leaks that cause prolonged pressure build-up are out-of-service items. According to the Commercial Vehicle Safety Alliance, a vehicle with more than 20 percent of its air brakes defective (measured with the engine off, reservoir pressure at 80–90 psi, and the brakes fully applied) shall be declared out of service.

5. Check air system for leaks (brakes applied).

Check the air system for leaks with the brakes fully applied. Begin by starting the engine and bringing the air supply to system pressure. With the engine idling, make and hold a full brake application until pressure stabilizes. If the compressor is unable to maintain reservoir pressure of 80–90 psi with the engine at idle and the brakes fully applied, it is recommended that the truck be deadlined until repairs are completed. Shut off the engine, and have an assistant hold the brakes applied while you examine the truck for audible air leaks.

6. Drain air tanks; test one-way and double-check valves.

The supply tank feeds the primary and secondary reservoirs of the brake system. Each is pressure protected by means of a one-way check valve. One-way check valve operation can be verified by draining the supply tank and watching for pressure loss in the other reservoirs. One-way check valves are used variously throughout an air brake system to pressure-protect and isolate portions of the circuit.

Two-way (double) check valves play an important role as a safeguard in dual-circuit air brake systems. The typical two-way check valve is a T with two inlets and a single outlet. It outputs the larger of the two source pressures to the outlet port and blocks the lower pressure source; the valve will shuttle in case of a change in the source pressure value. In other words, it will always prioritize the higher source pressure. Two-way check valves provide a means of providing the primary circuit with secondary circuit air and vice versa in case of a circuit failure. In the event that both source pressures are equal, as would be the case in a properly functioning dual air brake circuit, the valve will prioritize the first source to act on it. This would usually be from the primary circuit.

7. Check low air pressure warning devices.

Before performing a pressure drop test on a heavy truck a technician should ensure the engine is stopped and the system is at full pressure. Then make and maintain a brake application. Allow pressure to stabilize for one minute, and then begin timing for two minutes while watching the dash gauge. Pressure drop for single vehicles is 4 psi within two minutes and for singles' combination vehicles, 6 psi within two minutes. System air build-up times are defined by federal legislation, specifically FMVSS 121. A common check performed by enforcement agencies requires that the supply circuit on a vehicle be capable of raising air system pressure from 85 to 100 psi in 25 seconds or less. Failure to achieve this build-up time indicates a worn compressor, defective compressor unloader assembly, supply circuit leakage, or a defective governor.

FMVSS 121 requires that a driver receive a visible alert when the system pressure drops below 60 psi; in most cases, this is accompanied by an audible alert, usually a buzzer. A low air pressure warning device is fitted to both the primary and secondary circuits. This is a simple electrical switch that can be plumbed anywhere into a system requiring monitoring. The switch is electrically closed whenever the air pressure being monitored is below 60 psi. When the air pressure value exceeds 60 psi, the switch opens.

Verifying the operation of a low air pressure warning switch can be done by pumping the service application valve until the system pressure drops to the trigger value. A dash-located gauge must monitor both the primary and secondary circuit air pressures. The required visible warning is usually a dash warning light. An optional audible alert is almost always used in conjunction with the visible alert.

8. Check spring brake inversion/emergency (spring) brake control valve, if equipped.

To check for proper functioning of the inversion (spring brake control) valve, the air system should be fully charged (minimum 100 psi), and the spring parking brakes must be released. Next, locate and drain the wet tank and then the primary axle service reservoir completely by means of the draincocks. The primary gauge needle will drop to 0 psi, indicating a pressure loss for that system, which is normal. The secondary gauge needle must remain at system pressure (minimum 100 psi) to start the check of the inversion valve. Depress the foot valve in a normal manner. The spring brakes on the rear axle(s) will apply as the air is exhausted from the brake air chambers. Release the foot valve, and the spring brakes will release as air is supplied to them from the secondary reservoir. If the spring brakes do not or only partially release, then the inversion valve has failed and should be replaced. Remember, an inversion valve is a failsafe device and does not operate when brakes are functioning properly.

9. Check tractor protection valve, if equipped.

The tractor protection valve protects the tractor air supply under a trailer breakaway condition or severe air leakage. Two air lines connect a tractor brake system with a trailer air circuit. The trailer supply line supplies the trailer with air for braking and any other pneumatic systems such as its suspension. The trailer service line is the service brake signal line. Both these lines are plumbed to the tractor protection valve. When the trailer supply dash valve (emergency) red hexagonal is in its off position, the tractor protection valve is also off. When the driver makes a service brake application, no air will exit the tractor protection valve service signal line to the trailer. When the trailer air supply valve is open, air will open the tractor protection valve and supply the trailer pneumatic systems with air. Then when the driver makes a service brake application, the service brake signal air will be transmitted through the tractor protection valve to apply its service brakes.

It should be noted that the trailer air supply controls the trailer park brakes. Whenever air supply is interrupted, such as when the trailer is being parked or unintentionally in a breakaway, the spring brakes will apply regardless of how much pressure is in the trailer air reservoirs.

The tractor protection valve is designed to isolate the tractor air system from that of the trailer at a predetermined value that ranges from 20 to 45 psi, depending on the system. It should be noted that at these pressures, the spring brakes on both the tractor and trailer would be partially applied.

10. Test air pressure build-up time.

A common check performed by enforcement agencies requires that the supply circuit on a vehicle be capable of raising air system pressure from 85 to 100 psi in 25 seconds or less. Failure to achieve this build-up time indicates a worn compressor, defective compressor unloader assembly, supply circuit leakage, or a defective governor.

11. Check condition and operation of hand brake (trailer) control valve, if equipped.

The trailer control valve is used to actuate the trailer service brakes on a trailer independently of the tractor service brakes. This valve is also known by the terms: trolley valve, broker brake, and spike. The source of air supplied to the trailer control valve is usually the secondary circuit. Brake application pressure air is modulated proportionally with control valve travel.

Two-way check valves are used in the application circuit of the trailer brakes. The two-way check valve is located downstream from both the trailer control valve and the treadle valve. In a typical system, the trailer application brake valve (trolley or spike) uses secondary circuit sourced air and the treadle valve trailer service signal uses primary circuit sourced air to activate the trailer service brakes. If a driver was using the spike to bring the vehicle to a stop and during this process an emergency occurred that required a panic application of the treadle valve, the two-way check valve would permit the source air to change from secondary circuit air from the trailer control valve to the higher source pressure value from the treadle valve.

Trailer control valve performance can be verified with an accurate air pressure gauge. The trailer is supplied with air by means of air hoses. Service, and trailer supply, air is sent to the tractor protection valve.

12. Perform antilock brake system (ABS) operational system self-test. Perform automatic traction control (ATC) operational system self-test, if equipped.

All current ABSs are electronically managed. A typical ABS adapts a standard air or hydraulic system for electronic management. All current systems are designed to default to non-managed operation in the event of an electrical or electronic malfunction. New trucks are equipped with ABS brakes and are mandatory.

An ABS consists of a means of monitoring the speed of each wheel, an electronic control unit (ECU) or ABS module to manage the system, and a modulator assembly to affect the outcomes of the ECU logic processing by modulating the pressures to each wheel. ABS brakes are especially effective at managing split-coefficient braking conditions. Split-coefficient braking occurs when road surface conditions (icy, wet, gravel, dry, etc.) differ from one wheel to another.

All electronic ABS are equipped with self-diagnostics. These can be accessed by flash or blink codes using the dash warning light or a scan tool (digital diagnostic reader) connected either at the ATA connector in the cab or directly to the ABS module. Each time an electronic ABS is powered up, a self-test is performed. This consists of testing the signal range in the wheel speed sensors and cycling the solenoids in the system actuators. The latter test is audible, producing a clicking noise.

Access the ABS self-diagnostics by selecting the correct message identifier (MID). Identify the failure mode indicator (FMI) that has triggered the fault code and repair the failure using the OEM service instructions.

13. Inspect coupling air lines, holders, and gladhands.

Visually check all hoses and lines in the brake system. Service and trailer supply air is plumbed from the tractor protection valve. This air is transferred to the trailer by air hoses with couplers known as gladhands. Gladhands enable coupling between the tractor air supply and the trailer pneumatic circuit. Gladhand seals are usually manufactured out of rubber. They are retained in an annular groove in the gladhand and are easily replaced when they fail.

14. Check brake chambers and air lines for secure mountings, damage, and missing caging plugs.

Air brake chambers should be checked for signs of damage (see the following illustration). Visually check for damage from road hazards and look for cracks around the mounting studs. Check the air chamber bracket for cracks. Any hose or tubing that is worn, pinched, abraded, leaks, or swells under pressure should be replaced. Air lines should be routed to avoid damage from moving parts, and secured so that they are not damaged by an accumulation of ice during freezing weather.

Inlet port
Diaphragm
Pressure plate
Inlet port
Clamp ring nuts and bolts
Clamp ring
Return spring
Nonpressure plate
Push plate/rod assembly
Voke or clovis
Locknut

2013 © Delmar, Cengage Learning

Here are some of deficiencies that should be noted during an annual inspection:

- Absence of braking action on any axle, required to have brakes, upon application of the service brakes.
- Missing or broken mechanical components including: shoes, lining pads, springs, anchor pins, spiders, cam rollers, push rods, and air chamber mounting bolts.
- Loose brake components including air chambers, spiders, and camshaft support brackets.
- Audible air leak at brake chamber (ruptured diaphragm, loose chamber clamp, etc.).
- Hose with any damage extending through the outer reinforcement ply. Rubber impregnated fabric cover is not a reinforcement ply. Thermoplastic nylon may have braid reinforcement or color difference between cover and inner tube. Exposure of second color is cause for rejection.
- Bulge or swelling when air pressure is applied.
- Any audible leaks from hose or tubing.
- Two hoses improperly joined (such as a splice made by sliding the hose ends over a piece of tubing and clamping the hose to the tube).
- Air hose cracked, broken, or crimped.
- Tubing cracked, damaged by heat, broken, or crimped.

15. Inspect and record front and rear brake lining/pad condition and thickness.

Brake linings that have any of the following conditions should be replaced.

- Lining or pad is not firmly attached to the shoe
- Saturated with oil, grease, or brake fluid

- Non-steering axles: Lining with a thickness less than ¼ inches at the shoe center for air drum brakes, ¹⁄₁₆ inches or less at the shoe center for hydraulic and electric drum brakes, and less than ⅛ inches for air disc brakes.
- Steering axles: Lining with a thickness less than ¼ inches at the shoe center for drum brakes, less than ⅛ inches for air disc brakes, and ¹⁄₁₆ inches or less for hydraulic disc and electric brakes.

Unequal lining wear should be noted and the cause investigated.

Truck brake shoes are in most cases fixed-anchor assemblies mounted to the axle spider and actuated by an S-camshaft. Brake shoes for a heavy-duty truck can have their friction facings or linings mounted to the shoe by bonding, riveting, or by screw fasteners. Current application shoes are remanufactured and replaced as an assembly, that is, with the new friction facing already installed. When reusing shoes, check for arc deformities usually caused by prolonged operation without brake adjustment. Brake shoes must be fitted with the correct linings. Lining requirements can change with different vehicles and applications. When reconditioning shoes with bolted linings, if the original fasteners are reused, the lock washers should at least be replaced. Riveted linings should be riveted in the manufacturer's recommended sequence.

The friction rating of linings is coded by letter codes. Combination lining sets of shoes are occasionally used; these use different friction ratings on the primary and secondary shoes. When combination friction lining sets are used, care should be taken to install the lining blocks in the correct locations on the brake shoes.

It is good practice to replace all of the brake linings on a tandem drive axle truck on a PMI schedule. When the linings on a single wheel are damaged, as in the case of wet seal axle lube failure, the linings of both wheel assemblies on the axle should be replaced to maintain brake balance.

The friction pads in air disc brake assemblies should also be changed in paired sets, that is, both wheels on an axle. Vented disc brakes may have thicker inner pads than outer pads, while solid disc brakes usually have equal thickness inner and outer pads. Heat transfers more uniformly in the solid disc assemblies than vented ones, requiring thicker inboard pads.

16. Inspect condition of front and rear brake drums/rotors.

Brake drum condition can vary according to many different factors, including but not limited to, scoring, bell-mouth condition, concave-like, convex-like, threaded appearance, and hard spots. For example, convex condition indicates a drum diameter greater at the friction surface edges as compared to the center of the friction surface. Concave condition indicates a drum diameter greater at the center of the friction surfaces as compared to the edges. A bell-mouthed condition occurs when the drum diameter is greater at the inboard edge of the drum (next to the backing plate) compared to the outboard edge. The out-of-round condition is a variation in the drum diameter at measurements taken 180 degrees apart.

A Scored drum   B Bell-mouthed drum   C Concave drum

D Convex drum   E Hard spots   F Threaded drum

2013 © Delmar, Cengage Learning

When checking for lateral run out, a dial indicator is usually mounted on the hub and the face of the rotor is indicated. Total indicated run out should generally not be more than 0.005 inch (0.13 mm). If brake drums are to be reused, they should be inspected for wear, heat discoloration, scoring, glazing, threading, concaving, convexing, bell mouthing, and heat checking. It is not usually possible to machine a used brake drum that is warped due to heat tempering (recent practice and the low cost of drums usually results in drum replacement with a brake job). It is good practice to machine new drums before installation due to warpage caused by incorrect storage practices after manufacture. The maximum legal service limit for 14-inch-diameter brake drums is 14.120 inches. The machining limit is 14.060 inches and the discard limit is 14.090 inches. A drum micrometer is used to check brake drum diameter.

Brake rotors must be visually inspected for heat checking, scoring, and cracks. One must measure rotors for thickness with a micrometer. You use a dial indicator to check rotor run out and parallelism. In highway applications, rotors tend to outlast drums and are often reused after a brake job. They must be turned within legal service specifications using a heavy-duty rotor lathe.

"Long Stroke" Clamp Type Brake Chamber Data		
Type	Outside Diameter (inches)	Maximum Stroke at Which Brakes Must Be Readjusted (inches)
6	4½	1¼
9	5¼	1⅜
12	5¹¹⁄₁₆	1⅜
16	6⅜	1¾
20	6²⁵⁄₃₂	1¾
24	7⁷⁄₃₂	1¾
30	8³⁄₃₂	2
36	9	2¼

Bolt Type Brake Chamber Data		
Type	Outside Diameter (inches)	Maximum Stroke at Which Brakes Must Be Readjusted (inches)
16	$6\frac{3}{8}$	2.0
20	$6\frac{25}{32}$	2.0
24	$7\frac{7}{32}$	2.0
24*	$7\frac{7}{32}$	2.5
30	$8\frac{3}{32}$	2.5

*For 3 in. maximum stroke-type 24 chambers

Bolt Type Brake Chamber Data		
Type	Outside Diameter (inches)	Maximum Stroke at Which Brakes Must Be Readjusted (inches)
A	$6\frac{15}{16}$	$1\frac{3}{8}$
B	$9\frac{3}{16}$	$1\frac{3}{4}$
C	$8\frac{1}{16}$	$1\frac{3}{4}$
D	$5\frac{1}{4}$	$1\frac{1}{4}$
E	$6\frac{3}{16}$	$1\frac{3}{8}$
F	11	$2\frac{1}{4}$
G	$9\frac{7}{8}$	2

Wedge Brake Data		
Type	Outside Diameter (inches)	Maximum Stroke at Which Brakes Must Be Readjusted (inches)
9	$4\frac{9}{32}$	$1\frac{1}{2}$
12	$4\frac{13}{16}$	$1\frac{1}{2}$
16	$5\frac{13}{32}$	2
20	$5\frac{15}{16}$	2
24	$6\frac{13}{32}$	2
30	$7\frac{1}{16}$	$2\frac{1}{4}$
36	$7\frac{5}{8}$	$2\frac{3}{4}$
50	$8\frac{7}{8}$	3

Movement of the scribe mark on the lining shall not exceed $\frac{1}{16}$ inch.

17. Check operation and adjustment of front and rear brake automatic slack adjusters.

The slack adjuster is critical in maintaining the required freeplay and adjustment angle. Current slack adjusters are required to be automatically adjusting, but their operation must be verified routinely. Manually adjusting automatic slack adjusters is dangerous and should not be done, except during installation or in an emergency to move the vehicle to a repair facility, because manual adjustment of this brake component (1) fails to address the true reason why the brakes are not maintaining adjustment, giving the operator a false sense of security about the effectiveness of the brakes, which are likely to go out of adjustment again soon; and (2) causes abnormal wear to the internal adjusting mechanism for most automatic slack adjusters, which may lead to failure of this brake component.

Slack adjusters connect the brake chambers with the foundation assemblies on each wheel. They are connected to the pushrod of the brake chamber by means of a clevis yoke (threaded to the pushrod) and pin. Slack adjusters are spline mounted to the S-cams and positioned by shims and an external snap ring.

The slack adjuster converts the linear force of the brake chamber rod into rotary force or torque and multiplies it. The distance between the slack adjuster S-cam axis and the clevis pin axis will define the leverage factor; the greater this distance, the greater the leverage.

The objective of brake adjustment, whether manual or automatic, is to maintain a specified drum to lining clearance and a specified amount of freeplay. Freeplay is the amount of slack adjuster stroke that occurs before the linings contact the drum. Grease or automatic lubing systems lubricate slack adjusters. The seals in slack adjusters are always installed with the lip angle facing outward. When grease is pumped into the slack adjuster, grease will easily exit the seal lip when the internal lubrication circuit has been charged.

Failure to adjust the slack adjuster clevis (see the following illustration) with the proper template causes improper self-adjusting action in the slack adjuster. Observe the outer clevis pin hole through the slot in the template. If necessary, adjust the clevis on the end of the pushrod until the outer hole in the clevis is completely visible through the template opening.

With the clevis properly adjusted, the pushrod must extend at least ½ inch (1.27 mm) into the clevis threads, and not more than ⅛ inch (3.17 mm) into the yoke (see the following illustration).

Rotate the adjusting nut on the automatic slack adjuster so the openings in the slack adjuster arm are aligned with the openings in the clevis. Place anti-seize compound on the two clevis pins and insert these pins through the clevis and slack adjuster arm. When replacing a failed slack adjuster, the distance between the axis of the S-cam bore and the clevis pin bore must be maintained—a difference of ½ inch can greatly alter the brake torque and unbalance the brakes.

Brake Chamber Free Stroke Measurement

Follow this procedure to perform the free stroke measurement.

- Measure the distance from the brake chamber face to the center of the large clevis pin with the brakes released. Record this measurement.

- Insert a pry bar in the slack adjuster clevis and extend the brake chamber pushrod until the shoes contact the drum. Measure the distance between the center of the large clevis pin and the brake chamber face. The difference between these two measurements is the free stroke.

- If the free stroke is not within specifications adjust the slack adjuster (adjusting) nut to obtain the specified free stroke.

Brake Chamber Applied Stroke Measurement

- Measure the distance from the center of the large clevis pin to the brake chamber with the brakes released (see the following illustration). Record this measurement.

Measure "Free Stroke"

Free stroke = "Y" minus "X"
Drum brake free stroke must be 1/2–5/8 inches.
(12.7–15.9 mm)
Disc brake free stroke must be 3/4–7/8 inches.
(19.1–22.2 mm)

© 2013 Delmar, Cengage Learning

■ Have a co-worker apply the brakes until the application pressure is 80–90 psi (551.6–620.55 kPa). Measure the distance between the center of the large clevis pin and the brake chamber face. The difference between these two measurements is the applied stroke (see the following illustration).

| Measure while brake is released | Apply brakes with 80–90 psi | Measure while brake is applied |

© 2013 Delmar, Cengage Learning

■ If the applied stroke is not within specifications, adjust the slack adjuster (adjusting) nut to obtain the specified applied stroke.

▨ 18. Check S-camshaft and bushing condition.

When performing foundation brake servicing, all of the hardware mounted on the axle spider should be inspected and replaced if required. There should be no radial play of the S-camshaft. Rollers and cams should be inspected for flat spots, and it is good practice to replace any spring-steel components such as retraction springs and snap rings. Spider fastener integrity should be checked at each brake job. Spiders should be inspected for cracks at each brake job.

S-cam bushing seals should be installed so that the spider outboard bushing sealing lip faces inward toward the bushing and the inboard bushing sealing lip faces outward from the bushing. This allows excess grease to exit from the inboard side of the bushing and prevent it from being pumped into the foundation assembly. S-cam profiles depend on friction and should never be lubricated.

19. Lubricate all air brake component grease fittings.

If a slack adjuster passes inspection, the following procedure for lubrication should be followed. Apply a thin film of chassis grease to the slack-adjuster splines. After reassembly, pressure-lubricate the slack adjuster according to the manufacturer's instructions. Pressure-lubricate the in-board S-cam bushing until grease flows out of the slack adjuster end of the tube. Grease should not flow out of the out-board end of the tube. If it does, the seal is defective and it must be replaced. Also, do not replace the existing grease fitting(s) with the pressure relief type. Only standard non-vented fittings are to be used with spring loaded lip seals.

20. Check master cylinder for leaks and damage; check fluid level and condition.

When checking hydraulic brake fluid in a vehicle, not only should the level be checked, but any leaks from the master cylinder, reservoir, or brake line fittings as well. The appearance of the fluid is also an important part of the inspection. A cloudy or opaque look to the fluid will indicate some sort of contamination (dirt, rust, etc.), especially if severe. Fluid that is cloudy or darkened should be flushed and refilled with fresh fluid.

21. Inspect hydraulic brake lines, fittings, flexible hoses, and valves for leaks and damage.

A proper inspection of a heavy truck hydraulic braking system should include a visual checking around the wheel cylinders for an accumulation of contamination, which generally indicates leakage. Brake lines should be in good condition with no corrosion or damage that may compromise system performance. A thorough inspection of brake lines is always considered an important part in inspecting the hydraulic brake system.

22. Check hydraulic parking brake operation; inspect parking brake application and holding devices.

A check of the parking brake controls should come under an inspection of the hydraulic brake system. Test actuator freeplay and the parking brakes ability to effectively stop the vehicle. See the following illustration.

23. Check operation of hydraulic system; pedal travel, pedal effort, and pedal feel (drift).

The brake pedal should feel firm with minimal free travel at the top of the stroke. If the pedal feels spongy, it is usually an indication of air in the hydraulic system. Trucks with hydraulic brakes manufactured after 1976 are equipped with a tandem master cylinder, and dual circuit brakes (see the following illustration). In a dual circuit brake system, when one circuit fails, the remaining circuit will be unaffected. When there is a failure in one circuit, the brake pedal will have excessive free travel. A dash warning light should also be illuminated.

Dual Master Cylinder

Front brakes (disc)

Rear brakes (drum)

To warning light

Pressure differential valve and warning light switch

Metering valve

Proportioning valve

Three-way combination valve

2013 © Delmar, Cengage Learning

If the pedal slowly drifts to the floor there may be an internal leak in the master cylinder, or an external fluid leak. When the pedal is depressed, the brakes should apply promptly without excessive effort. If excessive pedal effort is required to stop the truck, the power brake booster should be checked. When the pedal is released, the brakes should release promptly. When the brakes are applied, all of the wheels should brake evenly and the truck should not pull to either side. If there is a pull, or one wheel locks up, or skids, the problem should be located and repaired before the truck is dispatched.

24. Inspect calipers and/or wheel cylinders for leakage and damage.

A proper inspection of a heavy truck's hydraulic braking system should always include a visual checking around the wheel cylinders for an accumulation of contamination, which generally indicates leakage. Look for evidence of leaks on the inboard side of the tire and wheel. When the drums are removed as part of a PMI, visually check the wheel cylinders by peeling back the rubber boot and checking for fluid. Leaking calipers may be examined with the wheel removed. Leakage from wheel cylinders can result in brake failure. Leaking wheel cylinders must be reconditioned or replaced.

25. Inspect power brake booster(s), hoses, and control valves.

Most trucks are equipped with power brakes to provide shorter stopping distances with less effort. A vacuum booster, a pneumatic booster, or a hydraulic booster may provide power brake assist. Hydraulic boosters are more effective and more common. When road testing the truck, it should stop without excess pedal effort. There should also be reserve capacity, providing for at least one power-assisted brake application after the engine is off.

The hydro-max power brake booster accomplishes this by engaging an electro-hydraulic pump when there is an interruption of fluid flow from the engine-driven hydraulic pump. The hydraulic power source may be the power steering pump, or may be a dedicated hydraulic pump. Vacuum-assisted systems have a vacuum reservoir that provides reserve capacity. Trucks with diesel engines and vacuum boosters need an engine-driven vacuum pump, because a diesel engine has pressure, not vacuum in the intake manifold. Air-over-hydraulic systems use air pressure for power assist, and the air in the tanks provides reserve capacity. Inspect the hydraulic system for fluid leaks. Check the hoses for signs of leaks, and rubber deterioration.

There are a number of control valves required for braking balance with disc/drum combinations. The metering valve delays application of front disc brakes until the rear drum brakes are ready to apply. If only the front brakes lock up or skid on slick surfaces when the brakes are lightly applied, the metering valve should be checked. You can test the metering valve by installing pressure gauges on the front and rear brake circuits. There should be no pressure at the front until the pressure at the rear reaches about 30–50 psi.

A proportioning valve is needed to reduce rear wheel lock-up during moderate to heavy brake applications. Some trucks are equipped with load sensing proportioning valves, which allow more pressure to the rear wheels as the frame is lowered by heavier loads. The rear wheels should not lock up before the front. There should be balanced braking at each of the truck's wheels. A pressure differential valve turns on a dash warning light when a loss of pressure occurs in one half of a split hydraulic brake system. Combination valves combine the metering, proportioning, and pressure differential valves into one assembly, which is replaced as a unit.

26. Check and/or adjust hydraulic drum brakes.

Adjust hydraulic drum brakes as follows:

- Remove the backing plugs for measuring the shoe-to-drum clearance. Be sure that, if used, the parking brake chambers are caged.
- Apply and release the brake pedal several times to center the brake shoes.
- Insert a feeler gauge through the backing plate opening and measure the clearance between the brake shoe and the drum. Typically, this clearance should be 0.040 to 0.100 inch (1.01 to 2.54 mm).
- If the shoe clearance is not correct, turn the manual adjusting wheel (star wheel) on the wheel cylinder to obtain the correct clearance.

27. Check operation of hydraulic assist back-up system and warning devices.

Trucks equipped with hydraulic-assist power brakes (usually hydraulic pressure is supplied from the power steering pump) require back-up to supplement power braking in the event the engine stops while the truck is moving. This is accomplished by a flow switch, which senses loss of hydraulic pressure from the power steering pump and actuates a 12-volt electro-hydraulic pump to provide emergency power assist. On some systems this is tested by making several brake applications with the key on and the engine off. When you have made enough pedal applications to exhaust the pressure in the accumulator, you should get a warning light and buzzer, and hear the 12-volt electro-hydraulic pump engage. Other systems are controlled by a solid state alarm/brake booster module and can verify when they are operating normally by observing both a primary and a secondary brake light and a tone alarm. This system is necessary for safety. It should be checked during each inspection.

2. Drivetrain (4 Questions)

1. Check operation of release/throw out bearing and clutch brake.

The clutch brake is a circular disc with a friction surface that is mounted on the transmission input spline shaft between the release bearing and the transmission. Its purpose is to slow or stop the transmission input shaft from rotating in order to allow gears to be engaged without clashing and to keep transmission gear damage to a minimum. Clutch brakes are used only on vehicles with non-synchronized transmissions. Worn clutch brake friction surfaces can lead to clashing (grinding) when moving the gear lever into reverse or low gear while vehicle is stationary.

To check the operation of the clutch brake, fully depress the clutch pedal in the cab and note the point of resistance that occurs before the end of the pedal stroke. Clutch brake squeeze should measure 1 inch at the clutch pedal. Most manufacturers recommend special tools for setting clutch brake. Check that the clutch brake friction faces are grease-free and the drive tabs are free.

2. Check clutch linkage/cable and levers for looseness or binding; lubricate release/throwout bearing as required.

When you service the release bearing on the clutch, all pivot points on the clutch linkage should be checked for free movement, inspected, and lubricated. On a push-type clutch, adjust clutch-pedal free travel. On pull-type clutches with a clutch brake, adjustment must accommodate proper free travel and clutch brake settings. Note that pull-type clutch internal adjustments must be completed prior to any external linkage adjustment. External linkage adjustments are usually not necessary.

3. Check clutch master cylinder fluid level; check clutch master cylinder, slave cylinder, lines, and hoses for leaks and damage.

Trucks equipped with a hydraulic clutch master and slave cylinder arrangement should have the master cylinder fluid level checked each time it is serviced. When fluid must be frequently added, it indicates that there is a leak in the system. Most leaks are found with a visual inspection of the cylinders, lines, and fittings. Brake fluid is most often used in clutch hydraulic systems. Hydraulic clutch master cylinders use DOT 3 brake fluid. The rubber seals in the hydraulic clutch system are incompatible with other fluids such as motor oil, power steering fluid, and hydraulic oils. Clutch master cylinders are usually not repairable and must be replaced.

4. Check and/or adjust clutch.

Apply the clutch pedal. Free travel should measure between 1 and 2 inches, with 1½ inches being typical. Measure the distance the pedal can be pushed down before thrust is exerted on the clutch release bearing to check clutch-pedal free travel.

Riding the clutch pedal means operating the vehicle with the clutch partially engaged. This can be destructive to the clutch as it permits slippage and generates excessive heat. Riding the clutch also puts constant thrust load on the release bearing, which can thin out the lubricant and cause excessive wear on the pads. Release bearing failures are often the result of this driving practice.

Both push-and pull-type clutches are disengaged through the movement of a release bearing. The release bearing is a unit within the clutch that mounts on the transmission input shaft but does not rotate with it. A fork attached to the clutch-pedal linkage controls the movement of the release bearing.

As the release bearing moves, it forces the pressure plate away from the clutch disc assembly. Manually adjusted clutches have an adjusting ring that permits the clutch to be adjusted to compensate for friction lining wear. The ring is positioned behind the pressure plate and threaded into the clutch cover. A lock plate secures the ring so that it cannot move. Levers are seated in the ring. When the lock strap is removed, the adjusting ring is rotated in the cover so that it moves toward the flywheel.

The wear compensator is a replaceable component that automatically adjusts for facing wear each time the clutch is actuated. Once facing wear exceeds a predetermined amount, the wear compensator allows the adjusting ring to be advanced toward the flywheel, maintaining pressure plate to clutch disc clearance within proper operating specification. This also keeps free pedal adjustment within specification.

The following five-step procedure can be used to determine if adjustment is required on manual and self-adjusting clutches with a clutch brake.

1. Remove the inspection cover.
2. With the clutch engaged (pedal up), measure the clearance between the release bearing housing and the clutch brake. This is the release travel. If the clearance is less than ½ inch or greater than 9/16 inch, continue with steps 3 and 4; if less than ½ inch or greater than 9/16 inch, go to step 5.
3. Release the clutch (pedal down) by fully depressing the clutch pedal.
4. Using the internal adjustment procedures for lock-plate and auto-adjust mechanisms, advance the adjusting ring until a distance of ½ inch to 9/16 inch is attained between the release bearing housing and the clutch brake (clutch cover) with the clutch pedal released (pedal down).
5. If the clearance between the release bearing housing and the clutch brake (clutch cover) is less than specifications, rotate the adjusting ring counterclockwise to move the release bearing toward the flywheel.

5. Check transmission and/or transfer case housing, fasteners, seals, filter, cooler, and cooler lines for cracks, leaks, and proper routing, if equipped.

The transmission case must be checked for any signs of fatigue. Cracking is a symptom that is usually accompanied by fluid leakage. Inspect all external connection points for evidence of leakage.

6. Check transmission wiring, connectors, seals, and harnesses for damage and proper routing.

The transmission wiring is essential to the operation of the truck. Other chassis computerized systems depend upon the transmissions tailshaft sensor for road speed data. Many new transmissions have computer controls, which are multiplexed with the rest of the truck electronics. A careful examination of the transmission's electrical components is important when performing preventive maintenance. Transmission wiring should be routed and secured so moving parts or hot exhaust pipes do not damage it. Connectors should be checked for external damage. Avoid disconnecting them during each inspection because they are designed to be weatherproof, and frequent disconnections can damage them. Make sure they are clean and dry.

7. Inspect transmission breather, service as needed.

Plugged breathers are associated with fluid leakage but not at the location of the breather. When a breather becomes plugged, fluid is often forced past seals or gaskets in the transmission. If a technician jumps to a conclusion when they notice fluid leakage at a seal, they may replace the seal and think the problem is solved. The technician should check the transmission breathers during any transmission diagnosis and PMI.

8. Inspect transmission mounts for looseness and deterioration.

Transmission mounts and insulators play an important role in preventing drive train vibration from transferring to the chassis of the vehicle. If the vibration were allowed to transmit to the chassis of the vehicle, the life of the vehicle would be reduced. Driving comfort is another reason why insulators are used in transmissions. Insulators absorb shock and torque. If the transmission was mounted directly to a rigid frame, the driveline torque required to haul heavy loads would need to be absorbed by the transmission resulting in shorter service life. Broken transmission mounts are not readily identifiable by any specific symptoms. They should be visually inspected during each PMI.

9. Check transmission oil/fluid level and condition.

Check transmission oil by removing the fill plug; oil should be exactly at the level of the fill hole or slightly above. During a PMI, fill the transmission to the correct level exactly. Do not overfill. This can cause aeration.

Cold temperature operations can cause transmission problems. High-viscosity oils typically used in transmission and axles thicken and may not flow at all when the vehicle is first started. Rotating gears can push lubricant aside, leaving voids or channels where no lubricant is actually contacting the gears. This is referred to as channeling. It is not until heat generated from under-lubricated gear contact surfaces melts the gelled lube that it starts to flow and do its job.

10. Inspect U-joints, yokes, driveshafts, and center bearings for looseness, damage, and proper phasing.

Loose end yokes, excessive radial runout, slip spline slop, bent shaft tubing, or missing plugs in the slip yoke cause vibrations, U-joint, and center bearing problems. A PMI will normally include the inspection of all of these, plus checking the center bearing for damage and the trunnions of the U-joints for wear. When inspecting drivelines, also check the driveshaft(s).

Often, vibration is too quickly attributed to the driveshaft. Before condemning the driveshaft as the cause of vibration, the vehicle should be thoroughly road-tested to isolate the vibration cause. To assist in finding the source, ask the operator to determine what, where, and when the vibration is encountered. It is very helpful to keep in mind some of the causes of driveline vibration. U-joints are the most common source if the vibration is coming from the driveshaft, while driveshaft balancing is the next most common. Pay special attention to phasing when removing or installing a driveshaft.

Center support bearings are used when the distance between the transmission and the rear axle is too great to span with a single driveshaft. The center bearing is fastened to the frame and aligns the two connecting driveshafts. It consists of a stamped steel bracket used to align and fasten the bearing to the frame. A rubber mount inside the bracket surrounds the bearing, which supports the driveshaft. The truck manufacturer's recommended driveshaft rating should always be adhered to when replacing a driveshaft. Oversized or undersized driveline components will lead to frequent driveline failures. Vibrations at high speeds are often indicative of a driveline problem.

11. Inspect axle housing(s) for cracks and leaks.

A technician should be able to identify potential drive axle problems. Incorrect lubrication will greatly affect the life of bearings, gears, and thrust washers. A lubrication problem will affect, and often cause, fatigue on drive train components including the drive axles.

Identify the root cause of any external leakage. Clean the housing thoroughly to remove accumulations of grime and dirt. Plug the breather and any other openings and steam clean or pressure-wash the axle assembly. Visually inspect for loose or missing fasteners, for stripped cross-threaded holes and for cracks in the housing. Look for cracks wherever there is a weld, such as around the flange where the spider is attached, and where the center housing is attached. Pay particular attention to suspension attaching points and spring saddle mounts, because these are high-stress areas. Cracks can be identified with dye and UV light.

12. Inspect axle breather(s); service as needed.

Investigate any source of leaks found on the axle differential. Replacing a seal if the axle differential breather is plugged does not cure leaking seals. A plugged breather produces high axle housing pressure, which can result in leakage past seals. Clean the breather by soaking in solvent and blowing it out with compressed air each time the lube level is checked.

13. Lubricate all drivetrain grease fittings.

The slip joint should be greased until grease is forced out of the relief hole. Then the hole should be covered until grease begins to ooze out around the seal. Clean zerk nipples before greasing to avoid contamination. Lubricate the cross and bearings of U-joints until the lube appears at all four seals. This flushes abrasive contaminants from each bearing assembly and assures all four are purged.

14. Check drive axle(s) oil level.

Drive axle oil level is checked in the same way transmission oil level is checked. The oil should be exactly at the level of the fill hole or slightly over. Never overfill: this can cause aeration. Ensure that EP (extreme pressure) lubricant is used in units using hypoid gearing. Try not to mix lubricants with different specifications.

15. Change drive axle(s) oil and filter; check and clean magnetic plugs.

A technician can sometimes determine the condition of a drive axle differential by the fluid. Most drive axles are equipped with a magnetic plug that is designed to attract metal particles suspended in the gear oil. A nominal amount of "glitter" is normal, but excess "glitter" indicates a problem requiring further investigation.

When metal particles are found on the magnetic drain plug, the technician should inform the customer of the condition and instruct the customer to monitor the condition and change oil in 30 days for reevaluation in case more serious problems are developing in the drive axle.

Oil change intervals vary by manufacturer, type of lubricant used, and operating conditions. Always follow the manufacturer's recommendations and use the specified lubricant.

Draining the lubricant while the unit is still warm ensures that any contaminants are still suspended in the lubricant. Flush the axle with clean axle lubricant of the same viscosity as used in service. Do not flush axles with solvents such as kerosene. Avoid mixing lubricants of a different viscosity or oils made by different manufacturers. Ensure that differentials with hypoid gearing are filled with an EP rated lubricant.

16. Check power take off (PTO) unit operation and mounting; check PTO U-joints and driveshaft for looseness, damage, and proper phasing, if equipped.

The PTO is an integral part of the transmission and should be serviced at the same intervals as the transmission. Check the PTO for leaks and loose mounting hardware. Check cable or lever connections for proper adjustment and lubricate moving parts. Lubricate PTO driveshaft U-Joints every time the chassis is lubricated. PTO gears can often be checked for wear or damage by removing the inspection or shift cover. If excessive wear or damage such as pitting, cracking, galling or missing gear teeth has occurred, the PTO will need to be rebuilt or replaced.

17. Change transmission oil/fluid and filters; check and clean magnetic plugs.

Most manufacturers suggest a specific grade and type of transmission oil, heavy-duty engine oil, or straight mineral oil, depending on the ambient air temperature during operation. Do not use multipurpose gear oil when operating temperatures are above 230°F (110°C). Many of these gear oils break down above 230°F (110°C) and coat seals, bearings, and gear with deposits that might cause premature failures. If these deposits are observed (especially on seal areas where they can cause oil leakage), change to heavy-duty engine oil or mineral gear oil to assure maximum component life.

Always follow the manufacturer's exact hydraulic fluid specifications. For example, several transmission manufacturers recommend DEXRON, DEXRON II, and type C-3 ATD-approved SAE 10W or SAE 30 oils for their automatic transmissions. Type C-3 fluids are the only fluids usually approved for use in off-highway applications. Type C-3 SAE 30 is specified for all applications where the ambient temperature is consistently above 86°F (30°C). Some, but not all, DEXRON II fluids also qualify as type C-3 fluids. If type C-3 fluids must be used, be sure all materials used in tubes, hoses, external filters, seals, and so on, are C-3 compatible.

18. Take transmission(s) oil sample.

Drain transmission oil. Halfway through the drain-off, fill the sample container. Correctly label and send for analysis.

19. Take drive axle(s) oil sample.

During a PMI, transmission and drive axle oil samples are taken for analysis of lubricating qualities, metal, and dirt content. Also, corrosion-causing agents such as sulfuric and hydrochloric acids are tested. For this reason, the transmission/drive axle oil should be at normal operating temperature to ensure that all possible contaminants and corrosives be contained in the oil in its normal operating condition.

3. Steering, Tires, and Wheels (4 Questions)

1. Check steering wheel and column operation for free play or binding.

The best method for checking a steering system is with the wheels on the floor and the engine running. Allow the engine to idle with the transmission in neutral and the parking brakes applied. While someone turns the steering wheel one-quarter turn in each direction from the straight ahead position, observe all the pivoted joints on the tie rod ends and drag link and other steering systems-related components. This allows the technician to check the steering linkage pivots under load. Turn the steering wheel back and forth while visually checking for looseness. Manual steering wheel play should not exceed 2 inches and power steering wheel play should not exceed 2 ½ inches.

A loose steering gear mount could be binding the steering column and cause the power steering to become increasingly harder to turn. Elongated mounting holes in the power steering pump bracket may cause a noise while in operation. Worn holes in the power steering pump-mounting bracket could cause premature belt wear. If leaks occur at any of the seal locations, replace the seal. When a leak is present at the high-pressure fitting, first tighten it to the specified torque. If this fails, replace the o-ring at this fitting.

2. Check power steering pump and hoses for leaks and mounting; check fluid level and condition.

Reservoir o-rings, driveshaft seals, high-pressure fittings, and dipstick caps are all possible leak sources. If leaks occur at any of the seal locations, replace the seal. When a leak is present at the high-pressure fitting, tighten it to the specified torque. If this fails, replace the o-ring or if necessary, the complete fitting. Inspect all routing to ensure that no line damage will occur due to contact with other objects. If a line is chafed but no leak is present, a replacement decision should be made depending on the severity and the probability of a future leak or breakdown.

Check fluid level at operating temperature of about 175°F.

3. Change power steering fluid and filter.

Most OEMs recommend the use of engine oil or automatic transmission fluid in power steering systems. Low fluid level will cause increased steering effort and erratic steering. It may also cause a growling or cavitation noise in the pump. Foaming in the remote reservoir may indicate air in the power steering system. Most OEM truck manufacturers recommend checking the power steering fluid level at operating or working temperature of 175°F.

With the engine at 1,000 rpm or less turn the steering wheel slowly from stop-to-stop several times to raise the fluid temperature. Check the reservoir for foaming as a sign of aerated fluid. The fluid level in the reservoir should be at the hot full mark on the dipstick. If the power steering fluid is contaminated with metal filings, improper fluid, or moisture, the system must be flushed. After the system is flushed, the filter in the remote reservoir must be replaced.

4. Inspect steering gear(s) or rack-and-pinion for leaks and mounting.

The steering gear must be mounted true to the frame or binding will result. Any evidence of leakage at the steering gear requires removal and repair.

5. Inspect steering shafts and U-joints condition and for proper phasing; inspect pinch bolts, splines, Pitman arm-to-steering sector shaft, drag link, tie rod ends, cross tube, and wheel stops.

Steering systems use U-joints to allow the shafts to turn on an angle. Loose steering-wheel-to-steering-shaft fasteners can cause excessive play and poor steering response.

Do not compress a column steering U-joint with a C-clamp or large pliers when testing a linkage joint for wear. This can compress the tension spring in the joint and give the impression that there is wear in the joint. Push against the joint with hand force. This should be sufficient to identify excessive wear.

A loose Pitman arm on the steering gear output shaft could cause steering shimmy. The Pitman arm is responsible for directional stability. A damaged Pitman arm can account for the steering wheel being off center. A loose sleeve clamp on the drag link adjuster may allow enough play in the sleeve to damage the adjusting threads on the steering arm, enough to change steering wheel position.

Tie rod clamp interference with the I-beam on a full turn can cause a noise when turning over bumps, a damaged tie rod assembly, and inaccurate camber measurement. A misadjusted drag link could cause the steering gear to operate off center and cause steering shimmy.

6. Check kingpin and thrust bearing wear.

Use the procedure indicated below for checking kingpin fit for play. A dial indicator is mounted as shown, and the top of the wheel is moved in and out. The indicator reads bushing wear and kingpin slop. Then reposition the dial indicator so that up-and-down movement will be indicated and test the thrust bearing by moving the tire up and down and watching the movement of the steering knuckle.

2013 © Delmar, Cengage Learning

7. Check front and rear wheel bearings/hub assemblies for looseness and noise, including extended service (sealed, close tolerance, and unitized) assemblies.

Always observe TMC wheel end procedure when adjusting wheel bearings. Bearing adjustment is very important on trucks equipped with ABS to ensure the correct wheel speed sensor air gap. To check bearing endplay, mount a dial indicator so that the indicator tip is contacting the hubcap-mounting surface, push the hub inward, zero the indicator, and pull the hub outward. Wheel bearing endplay should be 0.001 inch to 0.005 inch.

Defective rear wheel bearings produce growling noises at low speeds. A growling noise produced by a defective front wheel bearing is most noticeable while turning a corner. Rear wheel bearings usually do not require repacking because most are lubricated with rear-axle fluid. Never spin a dry wheel bearing with compressed air because it can damage critical contact surfaces.

8. Check oil level and condition in all non-drive axle hubs; check for leaks.

Hubs should be pre-lubed when installing the wheel assembly. After installation, the bearing and hub assembly must be filled to the indicated level in the calibrated inspection cover. Ensure that the rubber boot cap seals properly and that there are no cracks in the plastic calibration plate.

9. Adjust wheel bearings as needed (including one and two nut types) in accordance with manufacturers' specifications.

Wheel bearings should be inspected at each brake job and at each PMI service. Current trucks use wet bearings, that is, they are lubricated with liquid lubricant, usually gear lube. Bearings should be cleaned with solvent and air dried, ensuring that the cone is not spun out by the compressed air. The bearing should be inspected for spalling, galling, scoring, heat discoloration, and any sign of surface failure. When reinstalling wet bearings, they should be pre-lubed with the same oil to be used in the axle hub. When grease-packed bearings are used, the bearing cone must be packed with grease. This procedure may be performed by hand or using a grease gun and cone packer. Grease-lubricated wheel bearings require the use of a high-temperature axle grease. It is bad practice to pack wet bearings with axle grease and it may shorten bearing life by reducing lubrication efficiency.

Bearings fail primarily because of dirt contamination. This may be due to the conditions a vehicle is operated under or poor service practice. Lubrication failures caused by an inappropriate lubricant or lack of lubricant (caused by a failed wheel seal) also account for bearing failures; lack of lubricant will result in rapid failure and a bearing welded to an axle.

Maladjusted bearings fail rapidly, especially when preloaded. The result of high preload on a bearing can be to friction-weld it to the axle. Overloaded bearings tend to fail over time, not rapidly.

Most highway heavy-duty trucks use taper roller wheel bearings. The correct method of adjusting a taper roller wheel bearing is that outlined by the Truck Maintenance Council (TMC) division of the American Trucking Association (ATA). All bearing manufacturers in North America have endorsed this method, which supersedes any previous adjustment procedures outlined by them. The TMC bearing adjustment method requires that the bearing be seated to a specified torque value at the adjusting nut while rotating the wheel bearing; this is designed to seat the bearing with a preload.

Next, the adjusting should be backed off to between one-sixth and one-third of a turn to locate the adjusting nut to the jam mechanism. Finally, the bearing end-play must be measured with a dial indicator. The required specification is between 0.001 and 0.005 inches. End-play must be present. If at least 0.001-inch end-play is not present, the adjustment procedure should be repeated. The bearing setting is locked in place on the axle spindle by a lock or jam nut that should be torqued to specification after the adjustment procedure. In some axles, a split forged or castellated nut and cotter pin are used to retain the wheel assembly. Current recommended practice in the truck service industry is defined by the TMC and the above method was developed to establish a single standard to counter numerous highway wheel-off incidents.

10. Inspect tires for irregular wear patterns and proper mounting of directional tires.

Tires should be inspected when a "walkaround" is performed. Rub your hand across the tread to feel for irregular wear. The number and inboard/outboard location of heavy truck tires makes it imperative that a separate check of tires be made exclusive to a road test. Mismatched tires affect the drivability and stability of a vehicle; mismatching tires can cause handling and braking problems that seriously compromise safety. Many tires are unidirectional; check the tread geometry to ensure they are properly mounted.

11. Inspect tires for cuts, cracks, bulges, and sidewall damage.

Cuts in the tread and sidewall should be noted for safety purposes. Tires with cuts that show exposed tire cords must be replaced. Cracking can be caused by exposure to the harmful rays of the sun. Bulges in the tread area can indicate a broken belt below the tread surface (road test might indicate steering wheel side-to-side movement at low speeds or vibration at highway speeds). Sidewall damage (bulges, cuts) can lead to future tire failure.

12. Inspect valve caps and stems.

Perform a visual inspection and observe damaged, bent, or misaligned valve stems. Valve caps must be installed to prevent loss of air pressure. Salt, sand, and ice can cause air in the valve stem to discharge.

13. Measure and record tire tread depth; probe for imbedded debris.

Measure the tread depth at the inboard, center, and outboard tread grooves, and at several places around the circumference of the tire. Tires on the steer axle must have at least 4/32 inch tread remaining at the most worn spot. Tires on drive, trailer, and dolly axles must have at least 2/32 inch tread remaining. Look for small rocks and debris imbedded in the tread and remove any found. If left, they may burrow into the casing and cause a leak.

If any repair bulge is 3/8 inch above the tire surface, the tire must be replaced. If any tire has bumps, bulges, or cracks indicating separation of the tire plies, replace the tire. If retread tires have an opening at the edge of the tread, insert a probe in this opening. If the probe can be inserted more than a 1/2 inch into the opening, replace the tire.

14. Check and record air pressure; adjust as needed.

Periodic checking and record keeping of air pressure is the most effective means of prolonging tire wear. Check manufacturer service bulletins for recommended up-to-date air pressure settings. Air contained within the tire is what supports the entire weight of the vehicle. Stability, directional control, and braking are all affected by tire pressure.

15. Check for loose lugs and/or slipped wheels; check mounting hardware condition; service as needed.

Wheel nuts and lugs must be correctly torqued. Over or under-torquing of stud-piloted wheels can cause broken studs and cracked or loose wheels. The single flange nuts of hub-piloted wheels are less susceptible to this problem. Slipped wheels caused by loose lugs can be observed on split rim design if the paint on the rim is scuffed as the wheel drags during braking and rust marks emerge. Mounting hardware must be of the proper grade to provide the clamping and tensile strength necessary to hold wheels/rims in place and have the ability to stretch or bend without breaking. If any mounting studs are stretched or broken, then all on that wheel location must be replaced.

16. Re-torque lugs/nuts in accordance with manufacturers' specifications.

At recommended intervals re-torquing is necessary. This is due to the constant flexing of the stud/hub material. Overloading can be a factor. Manufacturers have specified intervals recommended for wheel and rim inspections.

17. Inspect wheels and spacers for cracks or damage.

Periodic maintenance intervals are important for the inspection of wheels and spacers. A visual inspection can reveal cracks in the rim area. To check for cracking in the wheel nut mounting area, sometimes the nut will have to be removed as some designs cover the mounting hole. Cracked wheels are usually caused by vehicle overloading. Spacers are designed to keep dual wheels from having the tires contact each other causing friction/overheat, chafing, etc. They must be correctly installed to keep wheels in alignment. Generally, spoke wheels experience greater alignment and balance problems than disc designs, but with proper installation and torquing, it is possible that spoke wheels can run virtually trouble-free.

18. Check tire matching (diameter and tread) on dual tire installations.

Tires of a different tread design mounted on the same axle are a deadline condition in many jurisdictions. You cannot mix radial tires with bias tires on a steering axle. Mixing tires on the rear axle is dangerous because radial tires expand more than bias tires. Matching tire sizes on dual wheels prevents tire tread wear from slippage due to uneven surface areas. Dual drive wheels may be measured with a tape measure, or caliper. Low-profile radial truck tires enhance the radial design to produce even lower costs per mile. The low-profile name comes from the tire's aspect ratio, which, for any tire, is calculated by dividing the tire's section height (tread center to bead plane) by its section width (sidewall to sidewall). Simply put, low-profile tires are "squatter" than conventional radials. Advantages offered by low-profile radials include lower weight (up to 10 percent less than standard radials), lower rolling resistance (again about 10 percent less), greater vehicle stability due to a lower center of gravity, a better footprint as a result of improved pressure distribution, high retreadability, improved fuel economy, better traction, and increased tread life.

19. Lubricate all steering grease fittings.

Lubricant specified by the manufacturer for the type of application and service interval must be used. Most manufacturers recommend that shop air pressure be regulated when applying grease to components. The normal suggested setting is 15 psi. This setting is recommended because the grease is required to purge the rust and moisture. If high pressure is used, then the grease merely bypasses the debris and does not provide a cleaning action. Steering gear shafts commonly use universal joints. These also require periodic lubrication. Binding of the steering wheel while turning could be caused by binding universal joints in the steering shaft. Steering not returning to center could be caused by lack of lubricant on the pivot points.

4. Suspension, Frame, and 5th Wheel (4 Questions)

1. Inspect front and rear suspension components (springs, hangers, shackles, spring U-bolts, insulators, radius rods, torque rods, load pads, walking beams, and equalizers); check U-bolt torque in accordance with manufacturers' specifications.

The leaf spring type suspension uses semi-elliptic leaf springs to cushion load and road shocks. The springs are mounted on saddle assemblies above the equalizer beams and are pivoted at the front end on spring pins and brackets. The rear ends of the springs have no rigid attachment to the spring brackets, but are free to move forward and backward to compensate for spring deflection. Springs should be inspected for cracks, sag, and surface wear. Multiple leaf spring assemblies should be replaced as an assembly.

2. Inspect shock absorbers for leaks and mounting.

When a vehicle hits a bump, the wheel and suspension move upward in relation to the chassis. This causes jounce and rebound. Shock absorbers dampen spring action from jounce and rebound, reduce body sway, and improve directional stability and driver comfort. Worn-out shock absorbers will allow the front end to bounce, causing steering wheel shake. Shock absorbers dampen spring oscillation and are always required on air spring suspensions. It is advisable to replace shock absorbers in pairs. Inspect shock absorbers with one end disconnected, for resistance, bent pistons, bushing failure, and bent piston sleeves.

3. Inspect air suspension components (air springs/bags, mounts, arms, hoses, valves, linkage, and fittings) for leaks and damage; check suspension ride height.

Air spring suspensions provide a smoother ride and automatically adjust ride height. A height control (leveling) valve controls the air supply to the air springs to maintain a constant ride height. These systems may have single- or dual-height control valves. The control and relay valves provide a means of deflating the air springs when uncoupling a trailer from a tractor. If the air springs were blown, the suspension would ride low. Misadjusted leveling (height control) valves in the rear air suspension system will cause a too-high condition. The height control valve lever should be parallel to the ground in the neutral position when the air suspension has been properly inflated to level the vehicle.

In a rear-axle air suspension system the air springs are mounted between the frame and the main suspension support beams and these beams are attached to the rear axle. Some trailer air suspension systems have a rigid beam on each side of the suspension. The front of this beam is retained in a bracket with a bushing. These systems use an eccentric pivot bolt in this bracket for axle alignment.

All suspension types have height specifications. Check the manufacturer's specifications regarding specific adjustment and measuring procedures. Usually spring types are measured with the vehicle not having any load installed (empty). With air spring suspension it usually will not make a difference as the height should not change loaded or not. Ride height is very important as it will impact on center of gravity, handling and driveline angles. Cab-over models require setting the suspension before clutch adjustment is attempted.

Most manufacturers recommend settings be made by deflating the air system and measuring the specification on the upward stroke of the fill cycle. It is also recommended to drive vehicle slowly onto a level shop floor, that brakes not be applied, chock wheels, and allow for suspension fore and aft movement while performing adjustments. Have air system pressure at minimum 100 psi. Confirm measurement height and place of measurement on the vehicle. Road test and reconfirm.

4. Inspect operation of tag/pusher axle and components for mounting and damage.

Inspect tag/pusher axle structural components (hangers, arms, axle and seats, airbags, air hoses, etc.) daily for damage, cracks, or excessive wear. The brakes should be checked daily before vehicle operation. Lubricate components on the chassis lubrication cycle.

5. Lubricate all suspension grease fittings.

Automatic chassis lube systems (ACLS) eliminate the need for hand lubrication and increase component life. An automatic chassis lube system consists of the following components. There must be a reservoir to hold the grease. An automatic timer regulates how often grease is distributed. This can be adjusted to meet operating conditions. There may be an air-driven pump actuated by a solenoid valve, or an electric motor-driven pump. Air-driven pumps use metering devices to dispense the correct amount of lube to each component. Other systems use an air or electric motor driven pump with multiple outlets that meter the correct amount of lube at the pump. Many current vehicles are now equipped with at least some permanently lubricated components. These do not have grease fittings installed.

6. Check tandem axle spacing.

It is not necessary to replace the axle-alignment washers when replacing a radius rod on a truck. If the washers are not worn beyond use, then they should be reused to ensure a correct realignment of the axle. Also, the number of washers and their respective positions (forward or rearward) should be noted and returned exactly as removed.

A proper directional thrustline occurs when a vehicle's rear wheels track directly behind the front wheels. This prevents dog tracking or abnormal Ackerman effect. Worn walking (equalizing) beam bushings can cause the rear axles to be out of alignment. Replacing a bent radius rod could correct axle alignment. An axle shaft replacement will not correct the alignment of the axle assembly. A defective tracking bar could cause an axle alignment problem. Turning the adjustment eccentric is a method of adjusting the rear-axle alignment. Rotating eccentric bushings in the spring hangers does adjust the rear position on some spring designs.

7. Inspect fifth wheel mount, bolts, slider, air lines, locks, pivot pins, bushings, and stops.

The fifth wheel (see the following illustration) is the standard semi-oscillating type that articulates on an axis perpendicular to the vehicle centerline. This design gives excellent stability and versatility while minimizing weight and space. The fifth wheel is bolted to the tractor frame by means of Grade 8 fasteners. Fastener integrity and torque should be checked at each PMI. The maximum allowable fifth wheel height is determined by subtracting the trailer height and tractor frame height from the maximum height of 13 feet, 6 inches. Many trailers are built with a 47-inch (119.38 cm) kingpin height. However, the trailer kingpin height must be measured to be sure of this height. The technician must remember the fifth wheel is designed to operate with the trailer level.

2013 © Delmar, Cengage Learning

8. Test operation of fifth wheel locking device; adjust as needed.

The fifth wheel couples the tractor to the trailer kingpin. During the tractor-to-trailer coupling process, the fifth wheel should be adjusted so the trailer bolster plate makes initial contact with the fifth wheel at a point 8 inches (20.32 cm) to the rear of the center on the fifth wheel mounting bracket. After the tractor is coupled to the trailer, the driver should get under the tractor and trailer and use a flashlight to visually inspect that the fifth wheel jaws are locked.

When checking fifth wheel adjustment, movement should not exceed ⅛-inch horizontal play at any time. The jaws should also be checked for ease of operation and the sliders should work smoothly. A sliding fifth wheel is designed to move forward or rearward on its mounting plate. This type of fifth wheel is mounted on tracks and locked in position. The locking mechanism may be released mechanically with a lever or by air pressure supplied to an air cylinder. Test the operation of either type.

9. Check mud flaps, brackets, and reflective devices.

These are required to be present. Check for damage and tears; replace if necessary.

10. Check pintle hook assembly, mounting bolts, and locks.

A draw bar is used to pull dollies and full trailers. A large steel ring on the front of the draw bar is coupled to a pintle hook on the back of the lead trailer. The pintle hook is bolted at the centerline of the lead trailer. A pintle hook has a spring-loaded latch that prevents the draw bar from disengaging. The pintle hook must be carefully examined to prevent accidents. Make sure the latch is in good condition and is securely mounted. Check for signs of wear or cracks. CVSA out-of-service criteria calls for deadlining if any welded repairs are visible, frame cross members are cracked in the attachment area, fasteners are missing or ineffective, or cracks are present anywhere in the pintle hook or draw bar assembly. No part of the draw bar eye should have any section reduced by more than 20 percent.

11. Clean and inspect fifth wheel plate for cracks and damage; lubricate fifth wheel plate and all grease fittings.

Grease should be applied to the pivots and smeared over the deck. Fifth wheel lubrication is very important to maintain proper fifth wheel life and proper operation. A dry fifth wheel top surface may cause jerky steering. The top surface of the fifth wheel must be lubricated with water-resistant lithium-based grease. Lubrication of the fifth wheel top plate may be done by direct application or through the grease fittings on the underside of the top plate. To distribute grease, most automatic systems use the same components. These components are as follows: a reservoir, which holds the grease supply until ready to be distributed; a pump, which delivers the lubricant through a network of grease lines; metering valves, which dispense grease; automatic timer mechanisms, which direct the flow of grease; and electric motors and/or other power sources (such as compressed air that pump it).

12. Inspect frame and frame members for cracks and damage.

To measure a truck frame, the vehicle should be parked on level ground, such as on a smooth floor or pavement. Measure the frame angle with a bubble protractor or other leveling device. "Down in the rear" indicates a positive frame angle. "Up in the rear" indicates a negative frame angle. Frame sideways occurs when one or both frame rails are bent inward or outward. The following are some causes: collision damage, fire damage, and using the truck for other than original design. Frame sag occurs when the frame rails are bent downward in relation to the rail ends. The following are some causes:

- Excessive loads
- Uneven weight distribution
- Holes drilled in frame flanges
- Too many holes drilled in the frame web
- Holes drilled too close together in the frame web
- Welding on the frame
- Cutting holes in the frame with a torch
- Cutting notches in the frame rails
- Fire
- Collision damage
- Using the truck for other than original design

Frame buckle occurs when one or both frame rails are bent upward in relation to the ends of the rails. The following can cause buckle: collision, using the truck for other than design intent, and fire.

Diamond-shaped frame occurs when one frame rail is pushed rearward in relation to the opposite frame rail. Vehicle tracking is affected by this condition. The following can result in diamond frame: collision, towing, or being towed with the chain only attached to one side of the truck frame.

Frame twist occurs when the end of one frame rail is bent upward or downward in relation to the opposite frame rail. The following causes can result in a twisted frame: collision damage, rollover, rough terrain operation, and uneven loading.

When fabricating frame reinforcements, taper the reinforcement plates to avoid abrupt changes in section modulus. Each reinforcement plate edge should have an angle less than 45 degrees.

13. Inspect body attaching hardware.

If the truck is equipped with a body such as a refrigerated or dry cargo body it will be attached to the frame with brackets, clamps, and bolts. The attaching hardware between the truck frame and the body should be checked during each inspection. Look for loose bolts, cracked brackets, and shiny places along the frame indicating that the body may be shifting.

14. Inspect cargo ramps. Inspect lift gates, cylinders, controls, hoses, wiring, and warning decals; check fluid level, service as needed.

Cargo ramps and lift gates are essential to safely moving freight. They should receive careful attention like all other safety devices. Cargo ramps should be checked for signs of cracking and fatigue. Lift gates should operate dependably and safely. Check the hydraulic pumps, cylinders, and hoses for signs of leakage and repair before a failure or injury occurs. The hydraulic reservoir should be checked during routine lubrication service. If fluid must be added frequently, the leak should be located and fixed as soon as possible. The lift gate electrical components should be examined for signs of corrosion and wear. Check the latches and safety chains for signs of weakness.

15. Inspect rear (ICC/DOT) impact guard.

On the rear of straight trucks and semi-trailers there are rubber bumpers that are used as cushions when backing into a loading dock. Vehicles manufactured after 1998 are required to be equipped with collapsible docking guards. These should be examined for damage during a PMI inspection since they are subject to constant abuse.

E. Road/Operational Test (3 Questions)

1. Check operation of clutch and gear shift.

Begin checking clutch operation by starting the engine and depressing the clutch pedal. Listen for abnormal noise. If there is noise when the pedal is depressed and it disappears when the pedal is released there may be a noisy release bearing or wear associated with the clutch intermediate plate. Attempt to put the transmission in gear. You should be able to do this without gear clash. Check the operation of the clutch brake or countershaft brake if gear clash occurs. With the transmission in gear and the brakes released slowly engage the clutch. The truck should begin moving smoothly. Observe any chatter, abnormal vibrations, or noise. Upshift and downshift in the normal manner through the gears and note any abnormal conditions for later diagnosis and repair. A pyrometer is used to indicate exhaust gas temperature and can provide the driver with engine loading information useful for shifting.

2. Check operation of all instruments, gauges, and lights.

Check that all instruments, gauges, and indicator lights function properly and indicate a reading on the instrument panel. Record on the road test report.

3. Check steering wheel for play, binding, and centering.

Test the operation of the steering wheel at both low and high speeds. Be sure the steering wheel is centered when driving the truck straight ahead. Note abnormalities.

4. Check operation of automatic or automated mechanical transmission.

During the drive test, note the operation of the automatic transmission. Upshifts should occur without over-revving the engine and engine RPMs should not over-run during a shift. The engine also should not lug down following a shift. Both upshifts and downshifts should be firm and should occur without harshness or shock to the drive train. If the truck is equipped with an automatic transmission temperature gauge, you should note its reading when the temperature has stabilized and compare it to factory specifications for your application.

5. Check road speed limiter.

Note the operation of the road speed governor. Reprogram if required.

6. Check cruise control.

During the road test, check hard cruise, soft cruise, and upper and lower droop. Reprogram if required.

7. Observe exhaust for excessive smoke.

The color and consistency of the exhaust can reveal combustion problems. When a diesel engine is running properly with a heavy load, exhaust emission should be clear. Note and report any smoking during the road test. It can be an indication of serious problems that could be avoided if addressed at a PMI.

8. Test service brakes.

The brakes that are activated by the brake pedal are referred to as the service brakes. Brake tests, either hydraulic or air type, have stopping distance criteria that must be attained. The FMVSS sets the standards used in the United States and Canada to define stopping distances as well as other braking safety criteria. Work the brakes during the road test using a mixture of snub braking and aggressive panic stops.

9. Verify engine/exhaust brake or retarder operation.

The periodic servicing and/or inspection procedures for exhaust brakes are simple to perform. Use the engine brake during the road test when the engine RPM is in torque rise. Note the disengage RPM. In multi-cylinder, internal compression brake systems verify engine brake operation in each cylinder and banks of cylinders. Engine brakes are not a deadline item but are likely to produce driver complaints if not functioning properly.

Trucks equipped with a retarder offer the same brake-saving advantages of an engine/exhaust brake, but the technology is different. Most retarders are incorporated into an automatic transmission. Caterpillar truck engines can be equipped with a brake saver mounted in the flywheel housing, which may be coupled with manual transmissions.

Retarders may be operated either manually, or automatically. During the drive test, check the retarder to ensure that it slows the truck down when applied. The sliding gate exhaust brake uses a pneumatically activated gate actuated by chassis system pressure. The air supply to close the gate is controlled by an electronically switched pilot valve. An aperture in the sliding gate permits minimal flow through the brake gate during engine braking. The butterfly valve version operates similarly.

10. Check operation of backup warning devices.

The backup warning indicator circuit should energize when the vehicle is placed into the reverse gear. An audible-pulse warning sound should be heard and the backup light illuminate. Backup alarms must function and are a deadline item in many jurisdictions.

INTRODUCTION

Included in this section are a series of six individual preparation exams that you can use to help determine your overall readiness to successfully pass the Preventive Maintenance Inspection (T8) ASE certification exam. Located in Section 7 of this book, you will find blank answer sheet forms you can use to designate your answers to each of the preparation exams. Using these blank forms will allow you to attempt each of the six individual exams multiple times without risk of viewing your prior responses.

PREPARATION EXAM 1

1. The fifth wheel to upper coupler connection should have no more than how many inches of longitudinal play?

    A. $1/16$

    B. $1/8$

    C. $1/4$

    D. $1/2$

2. When a starting motor is removed, there are shims positioned between the starting motor mounting flange and the flywheel housing. Technician A says these shims provide improved electrical contact between the starting motor and the flywheel housing. Technician B says these shims provide easier installation of the starting motor mounting bolts. Who is correct?

    A. A only

    B. B only

    C. Both A and B

    D. Neither A nor B

2013 © Delmar, Cengage Learning

3. What is the operation being performed in the illustration above?

    A. Checking wheel-hub runout
    B. Aligning wheel hub
    C. Installing wheel bearings
    D. Installing a wheel seal

4. Technician A says that the blower motor should operate in all speed positions. Technician B says idle shutdown duration is programmable on electronic engines. Who is correct?

    A. A only
    B. B only
    C. Both A and B
    D. Neither A nor B

5. Technician A says that you can mix radial and bias tires on a steering axle as long as the vehicle doesn't exceed 55 mph. Technician B says that tires of different tread design mounted on the same axle are a deadline condition in many jurisdictions. Who is correct?

    A. A only
    B. B only
    C. Both A and B
    D. Neither A nor B

6. If the fuel is contaminated with water, the water will:

    A. Remain mixed with the fuel.
    B. Float on top of the fuel.
    C. Settle to the bottom of the fuel tank.
    D. Flow through the filters into the injection system.

7. Technician A says the back-up warning system provides a warning to anyone standing behind the truck. Technician B says the back-up warning system provides a visual warning. Who is correct?

    A. A only
    B. B only
    C. Both A and B
    D. Neither A nor B

8. A binding steering wheel may be caused by:

    A. Loose tie rod ends
    B. Loose or worn steering shaft U-joints
    C. Loose drag-link ends
    D. Worn kingpin bushings

9. Technician A says that an inductive clamp ammeter could be used to check alternator output. Technician B says that alternator output decreases as temperature increases. Who is correct?

    A. A only
    B. B only
    C. Both A and B
    D. Neither A nor B

10. An audible air leak is observed that appears to be coming from the front brake service brake chamber with the brakes applied. Which of the following is the LEAST LIKELY cause?

    A. Loose mounting bolts
    B. Ruptured diaphragm
    C. Loose chamber clamp
    D. Missing clamp bolt

11. During a PMI, Technician A says the air passages through the air-to-air intercooler should be checked for restrictions. Technician B says the air-to-air intercooler uses refrigerant to lower the intake air temperature. Who is correct?

    A. A only
    B. B only
    C. Both A and B
    D. Neither A nor B

12. If a starter ground circuit test reveals a voltage drop of more than 0.5 volts, which of the following may be the cause of the problem?

    A. A shorted armature winding
    B. An open field coil
    C. A damaged battery ground cable
    D. A faulty ignition switch

13. The most likely cause of blue smoke in diesel exhaust is:

    A. Coolant entering the combustion chambers.

    B. A restricted intake.

    C. Low fuel pressure to the injectors.

    D. Oil entering the combustion chambers.

14. What does the gauge in the illustration above show?

    A. Water temperature

    B. Oil temperature

    C. Oil pressure

    D. Transmission temperature

15. When adjusting hydraulic drum brakes, Technician A says to use a pry bar to center the brake shoes. Technician B says to turn the manual adjusting wheel (star wheel) to adjust shoe clearance. Who is correct?

    A. A only

    B. B only

    C. Both A and B

    D. Neither A nor B

16. Technician A says drive axle oil samples are taken for analysis of lubricating qualities, metal, and dirt content. Technician B says that oil samples are taken to check for corrosion-causing agents. Who is correct?

    A. A only

    B. B only

    C. Both A and B

    D. Neither A nor B

17. A gasoline engine experiences rough idle operation. This problem is worse when the engine is at normal operating temperature. The cylinder compression is within specifications and all the ignition components are satisfactory. Which of the following could be the cause of this problem?

    A. An exhaust gas recirculation (EGR) valve stuck open

    B. A burned exhaust valve

    C. Valve adjustment that is too tight

    D. A bent intake valve

18. A turbocharger on a tractor has less than the specified boost pressure. All of these defects may be the cause of the problem EXCEPT:

    A.   Turbocharger bearings that are starting to fail.

    B.   Improper wastegate adjustment.

    C.   Damaged compressor wheel.

    D.   Turbocharger oil seals leaking oil into the exhaust.

19. A wheel cylinder:

    A.   Does not need any maintenance.

    B.   Needs to be inspected at a PM.

    C.   Needs to be replaced periodically.

    D.   Is usually constructed of aluminum.

20. Technician A says that it is OK to use ether on glow-plug–equipped engines. Technician B always uses ether when starting a diesel engine. Who is correct?

    A.   A only

    B.   B only

    C.   Both A and B

    D.   Neither A nor B

21. While discussing alternator output testing, Technician A says when the alternator is full-fielded, the voltage should be kept below 17 V. Technician B says the output may be tested by lowering the system voltage to 11.5 V. Who is correct?

    A.   A only

    B.   B only

    C.   Both A and B

    D.   Neither A nor B

22. What would occur if the circuit, indicated at point X in the illustration above, was found to be heavily corroded?

    A. The right headlight would be inoperative.

    B. The left headlight would be inoperative.

    C. The right high beam would be dim.

    D. The left high beam would be dim.

23. Technician A says that an inspection of the road speed governor is required during the road/operational test. Technician B says that more emphasis should be placed on the post-trip inspection rather than the pre-trip inspection. Who is correct?

    A. A only

    B. B only

    C. Both A and B

    D. Neither A nor B

24. When inspecting an oil filter housing or its mounting, which of the following is LEAST LIKELY to be a procedure?

   A.   Visually checking for cracks
   B.   Visually checking gasket surface for nicks
   C.   Magna fluxing the housing and metal parts to detect small cracks
   D.   Visually inspecting housing passageways for obstruction and cracks

25. What is the most likely mileage recommended in changing synthetic lubricant for truck drive axles driven in a linehaul application?

   A.   500,000 miles
   B.   200,000 miles
   C.   100,000 miles
   D.   50,000 miles

26. Technician A says that a vehicle must have proof of inspection according to the Federal Highway Administration. Technician B says that any wiper unit that has inoperable parts on the driver's side shall be cause for deadline according to CVSA safety standards. Who is correct?

   A.   A only
   B.   B only
   C.   Both A and B
   D.   Neither A nor B

27. What is not checked when inspecting an air-brake system?

   A.   Tractor protection valve
   B.   Spring brake inversion system
   C.   Desiccant pack in air dryer
   D.   Operation of the clutch brake

28. A heavy-duty diesel-powered truck is having its electrical system inspected. The interior lights glow dimly even at a high engine speed. Technician A says that a defective ballast resistor should be suspected. Technician B says to check the cab ground integrity. Who is correct?

   A.   A only
   B.   B only
   C.   Both A and B
   D.   Neither A nor B

To HR-1
relay valve

2013 © Delmar, Cengage Learning

29. What is indicated in the illustration above?

   A.   Hydraulic liftgate cylinder

   B.   By-pass solenoid for reefer unit

   C.   Remote automatic kingpin locking relay

   D.   Hydraulic spring parking brake actuator

30. While discussing water pumps, Technician A says a defective water pump bearing may cause a growling noise with the engine idling. Technician B says the water pump bearing may be ruined by coolant leaking past the water pump seal. Who is correct?

   A.   A only

   B.   B only

   C.   Both A and B

   D.   Neither A nor B

31. Cloudiness in hydraulic brake fluid is a sign of which of the following?

   A.   Air in the fluid

   B.   Dirt contamination

   C.   Overheated/burnt fluid

   D.   Cold fluid

32. A battery is being tested on a truck during PMI inspection. Technician A says that the surface charge should be removed from a battery before it can be load tested. Technician B says when load testing, you should apply a load equal to half of the battery's cold cranking ampere (CCA) rating for 15 seconds. Who is correct?

   A.   A only

   B.   B only

   C.   Both A and B

   D.   Neither A nor B

33. All of the following problems may cause a low, spongy brake pedal EXCEPT:

   A.   Excessive brake pedal freeplay.

   B.   Air in the hydraulic brake system.

   C.   Brake fluid contaminated with moisture.

   D.   Low brake fluid level.

34. What do cracks between wheel lug holes indicate?

    A. Over inflation of tires
    B. Overloading
    C. Using larger than recommended tire sizes
    D. Incorrect wheel size for application

35. Technician A says that when filling the cooling system, the air bleeder valves should be closed. Technician B says that if the coolant solution contains over 67 percent antifreeze, heat transfer from the cylinders to the coolant is adversely affected. Who is correct?

    A. A only
    B. B only
    C. Both A and B
    D. Neither A nor B

36. When adjusting the height control valve in an air suspension system:

    A. The height control valve arm may be moved upward to lower the suspension height.
    B. Steel locating pins are installed in the height control valve arms.
    C. The air brake system pressure must be 70 psi (483 kPa) or more.
    D. The truck must have an average load on the chassis.

37. Technician A says a loose grab handle may cause personal injury. Technician B says the steps must be level and free of debris. Who is correct?

    A. A only
    B. B only
    C. Both A and B
    D. Neither A nor B

38. On the hydraulic drum brake in the illustration above, the technician is:

    A. Performing a hydraulic drum brake adjustment.
    B. Changing a return spring on a hydraulic brake.
    C. Measuring brake shoe to drum clearance.
    D. Setting the park brake gap.

39. While discussing alternator noise complaints, Technician A says a loose alternator belt may cause a squealing noise. Technician B says a shorted alternator diode may cause a whining noise. Who is correct?

    A. A only
    B. B only
    C. Both A and B
    D. Neither A nor B

40. The S-cam on one foundation brake fitted with Q brakes rolls over. Which of the following is the most likely cause?

    A. Excessively worn drums and shoes
    B. Application air pressure too high
    C. Improperly installed slack adjuster
    D. Brake chamber pushed rod too long

41. Technician A says air seat spring suspensions eliminate the need for cab shock absorbers. Technician B says you can adjust the leveling valve to maintain the proper seat height. Who is correct?

    A. A only
    B. B only
    C. Both A and B
    D. Neither A nor B

42. The following are part of an automatic chassis lube system EXCEPT:

    A. Grease lines.
    B. Reservoir.
    C. Pump.
    D. Manifold or distribution block.

43. While discussing safety starting switches, Technician A says the safety starting switch is connected parallel to the solenoid windings. Technician B says a defective safety starting switch may cause higher than specified starter current draw. Who is correct?

    A. A only
    B. B only
    C. Both A and B
    D. Neither A nor B

44. What activates the pump in an air-driven automatic chassis lube system (ACLS)?

    A. A solenoid valve
    B. A relay valve
    C. An electronic control module (ECM)
    D. A toggle switch inside the cab

45. When troubleshooting a diesel engine which uses excessive amounts of oil, which of the following should be checked first?

    A. Worn oil rings
    B. Oil leakage through the engine air compressor
    C. External oil leaks
    D. Oil leakage through the turbocharger

46. The charging indicator light on a truck is illuminated when the ignition switch is turned off. Technician A says some of the alternator diodes may be defective. Technician B says the wire connected to the alternator L terminal may be grounded. Who is correct?

A. A only
B. B only
C. Both A and B
D. Neither A nor B

47. The machining limit of a brake drum is 14.060 inches, and the discard limit is:

A. 14.075 inches.
B. 14.080 inches.
C. 15.005 inches.
D. 14.090 inches.

48. Technician A says a transmission oil sample can be taken after the oil is drained into a drain pan. Technician B says to take a transmission oil sample mid-way through the drain-off. Who is correct?

A. A only
B. B only
C. Both A and B
D. Neither A nor B

49. Technician A says to check the oil level with the engine running. Technician B says an oil level too low can cause engine failure. Who is correct?

A. A only
B. B only
C. Both A and B
D. Neither A nor B

50. When servicing the power take off (PTO) unit, which of the following is LEAST LIKELY to result in replacement of the PTO?

A. Missing gear teeth
B. Leaks
C. Cracking
D. Galling

PREPARATION EXAM 2

1. All of the following may result in an undercharged battery EXCEPT:

    A. A defective voltage regulator causing a high voltage regulator setting.

    B. Excessive resistance between the alternator battery terminal and the positive battery terminal.

    C. Loose or worn alternator drive belt.

    D. A 60-ampere fuse link installed in place of a 100-ampere fuse link in the alternator battery wire.

2. A technician is testing the low air pressure warning system on a truck by performing a pressure drop test on a straight single truck's air brake system. Technician A says that the pressure drop should be no more than 6 psi in two minutes. Technician B says you hold down the brake pedal for the first minute then release it. Who is correct?

    A. A only

    B. B only

    C. Both A and B

    D. Neither A nor B

3. What is the most likely cited maintenance item on transport trailers?

    A. Tires

    B. Lights

    C. Suspension

    D. Brakes

4. Technician A says to connect the battery positive cable first when connecting a battery. Technician B says to disconnect the ground cable first when disconnecting a battery. Who is correct?

    A. A only

    B. B only

    C. Both A and B

    D. Neither A nor B

5. What should a technician do when testing the operation of the trailer control valve?

    A. Move the handle to the fully applied position and record the air pressure.

    B. Drain the air system to 0 psi, and then allow pressure to build up again.

    C. Record the gauge reading of the air pressure first, and then drain the system.

    D. Listen for air leakage around the handle as the handle is moved to the applied position.

2013 © Delmar, Cengage Learning

6. What is the component shown in the illustration above?

    A. A one-way clutch
    B. A vertical yoke support
    C. An exhaust pipe support
    D. A center bearing support

7. When checking the emission control systems of a gasoline-fueled truck, always test:

    A. The carburetor fuel inlet line.
    B. The positive crankcase ventilation (PCV) valve.
    C. The air cleaner vacuum hoses.
    D. The intake manifold service port.

8. A roadside inspection is performed while it is raining. All of the following defects would be reason to tag the vehicle out of service EXCEPT:

    A. The wiper blade on the driver's side is cracked and badly worn.
    B. The wiper motor is inoperative.
    C. The passenger's side wiper blade is cracked and badly worn.
    D. The driver's side wiper arm is distorted so the blade does not contact the windshield properly.

2013 © Delmar, Cengage Learning

9. In the illustration above, which test is being performed?

    A. A battery drain test
    B. A battery capacity test
    C. A battery open circuit voltage test
    D. A starter current draw test

10. Technician A says that multiple leaf springs should be replaced as an assembly. Technician B says that one end of semi-elliptic springs are free to move forward and backward. Who is correct?

    A. A only
    B. B only
    C. Both A and B
    D. Neither A nor B

11. Technician A says that the air cleaner element should be replaced when 25 in. Hg air restriction is indicated. Technician B says that regardless of regulations, a truck must be serviced in the most cost-effective manner to the owner. Who is correct?

    A. A only
    B. B only
    C. Both A and B
    D. Neither A nor B

12. When aiming headlights with a headlight aligning screen:

    A. The vehicle does not need to be parallel to the aligning screen.
    B. The tire inflation pressure does not affect headlight alignment.
    C. The headlight alignment may be changed by installing new sealed beams.
    D. The headlight alignment is changed by turning adjusting screws on the headlight assembly.

13. Technician A says that technicians that perform mobile refrigerant system service must be certified. Technician B always makes sure both hand valves on the manifold gauge set are closed before connecting the hoses to the vehicle. Who is correct?

    A. A only
    B. B only
    C. Both A and B
    D. Neither A nor B

14. How should a truck's engine mounts be dynamically checked?

    A. visually inspecting for cracks and/or deterioration
    B. correlating information from the driver and the vehicle's maintenance history
    C. observing the engine mounts while the engine is torqued with its brakes applied
    D. automatically deadlining a vehicle due to the driver's suspicions

15. While checking fan blades, Technician A says a cracked fan blade may cause a clicking or creaking noise with the engine idling. Technician B says you should not stand or position your body in line with the fan while increasing the engine speed. Who is correct?

    A. A only
    B. B only
    C. Both A and B
    D. Neither A nor B

2013 © Delmar, Cengage Learning

16. How would suspension height of the illustration above most likely be checked?

    A. Measure from the road surface to the lower plate.

    B. Measure between the upper and lower plates.

    C. Measure from the road surface to the upper plate.

    D. Measure the angle of the regulator valve arm.

17. Technician A says a 45-degree elbow in an air brake line may be replaced with a 90-degree elbow without affecting brake operation. Technician B says that if a trailer service system has an air fitting that is smaller than the original line, the air timing/balance is affected. Who is correct?

    A. A only

    B. B only

    C. Both A and B

    D. Neither A nor B

18. Which of the following does the push hydraulic circuit in a cab lift system do?

    A. Lowers the cab back from a fully extended position

    B. Raises the cab from fully extended to the 45-degree tilt position

    C. Allows the cab to descend to the 60-degree overhead position

    D. Raises the cab from the lowered position to the desired tilt location

19. When adjusting a clutch linkage, Technician A says that the pedal-free travel should be about 1½ to 2 inches (38.1 to 50.8 mm). Technician B says that release bearing to clutch brake travel should be less than ½ inch (12.7 mm). Who is correct?

    A. A only

    B. B only

    C. Both A and B

    D. Neither A nor B

20. Technician A says an ohmmeter should be used to check for a short circuit between two circuits. Technician B says the battery must be fully charged before checking a circuit for current draw. Who is correct?

    A. A only

    B. B only

    C. Both A and B

    D. Neither A nor B

21. What is the LEAST LIKELY cause of a vehicle wandering?

    A. Mismatched or under-inflated tires
    B. Unmatched spring design/load capacity spring assemblies
    C. One or more broken spring leaves
    D. Frame alignment within manufacturer's specifications

22. Technician A says that when checking the condition of hydraulic drum brakes on a vehicle, to look for leakage and contamination around the wheel cylinders. Technician B inspects the brake lines for evidence of corrosion. Who is correct?

    A. A only
    B. B only
    C. Both A and B
    D. Neither A nor B

23. While checking fuses in a tractor/trailer, the technician finds an open fuse. What is the next step?

    A. Replace the fuse with the next higher-amperage fuse.
    B. Check the affected circuit for a short to ground.
    C. Check the affected circuit for an open.
    D. Install a circuit breaker of the same amperage as the fuse.

24. The Federal Highway Administration has set up a minimum inspection standards program. Technician A says that if a vehicle decal is used, a copy of the inspection form must be kept on file and should indicate where an inspector can get a copy of it. Technician B says that a tractor/trailer is considered to be one vehicle, requiring only one decal. Who is correct?

    A. A only
    B. B only
    C. Both A and B
    D. Neither A nor B

25. Most vehicle manufacturers recommend checking the level of power steering fluid at which working temperature?

    A. 100°F
    B. 140°F
    C. 90°F
    D. 175°F

26. A high resistance is suspected in the starting motor ground circuit on a negative ground electrical system. Technician A says to connect the voltmeter leads from the starting motor housing to the positive battery terminal. Technician B says the fuel injection system should be disabled and the engine cranked to read the voltage drop. Who is correct?

    A. A only
    B. B only
    C. Both A and B
    D. Neither A nor B

27. Technician A says that a clutch brake needs to be inspected for wear and fatigue. Technician B says that providing the clutch brake drive lugs are intact, it will always function properly. Who is correct?

    A. A only
    B. B only
    C. Both A and B
    D. Neither A nor B

28. Wires for fusible links should be:

    A. The same gauge size as the circuit wire.
    B. The same color as the circuit wire.
    C. Two gauge sizes larger than the circuit wire.
    D. A smaller gauge size than the circuit wire.

29. All of the following could cause a drive axle failure of a single drive axle vehicle EXCEPT:

    A. Incorrect ratio.
    B. Normal wear.
    C. Lubrication.
    D. Fatigue.

30. A cooling system is being pressure tested. Technician A says that a radiator pressure cap reading of 15 psi is normal. Technician B says that it is necessary to warm the engine to normal operating temperature before conducting a pressure test. Who is correct?

    A. A only
    B. B only
    C. Both A and B
    D. Neither A nor B

31. The headlight switch has failed in the circuit shown in the illustration above, Technician A says the parking lights are not affected. Technician B says the stoplights will fail to operate. Who is correct?

    A. A only

    B. B only

    C. Both A and B

    D. Neither A nor B

32. Technician A says that trucks with hydraulic-assist power brakes are not required to have a back-up system. Technician B says that on systems that do incorporate a back- up that the system should be checked during each inspection. Who is correct?

    A. A only

    B. B only

    C. Both A and B

    D. Neither A nor B

33. Technician A says that checking clutch operation and transmission shifting is part of a drive test during a preventive maintenance inspection (PMI). Technician B claims that when a drive test is performed, road speed governor actual speed should be checked. Who is correct?

    A. A only

    B. B only

    C. Both A and B

    D. Neither A nor B

34. The tractor protection valve closes to protect the tractor air supply if the air pressure in the supply line drops below what psi value?

    A. 20 psi (138 kPa)
    B. 55 psi (379 kPa)
    C. 75 psi (517 kPa)
    D. 80 psi (552 kPa)

35. A heavy-duty truck is found to have an exhaust leak. Technician A intends to either repair it immediately or deadline the truck. Technician B says that the vehicle can only be put out of service if the leaking exhaust fumes enter the driver/sleeper compartment or if they are likely to result in charring, burning, or damaging the wiring, fuel supply, or other combustible parts. Who is correct?

    A. A only
    B. B only
    C. Both A and B
    D. Neither A nor B

36. An applied pressure drop test is being performed on a straight truck air brake system. Technician A says that the pressure drop should be no more than 6 psi in two minutes. Technician B says you hold down the brake pedal for the first minute, and then release it. Who is correct?

    A. A only
    B. B only
    C. Both A and B
    D. Neither A nor B

37. When discussing auxiliary power unit (APU) maintenance, all of the following are true EXCEPT:

    A. Battery powered systems are more maintenance-intensive than engine-based systems.
    B. On engine-based systems, specific points such as the radiator and condenser fins should be cleaned and/or inspected.
    C. On battery-powered systems, a visual inspection should be performed as well as checking batteries and connections.
    D. Many APU failures can be traced to environmental factors such as vibration, corrosive road spray, and extreme heat and cold.

38. Technician A says an engine oil sample should be taken during every PMI. Technician B says to take an engine oil sample mid-way through the sump drain-off. Who is correct?

    A. A only
    B. B only
    C. Both A and B
    D. Neither A nor B

39. Technician A says that a circle inspection includes items such as torn mud flaps, leaking hoses, and outdated permits. Technician B says to perform a circle inspection first before any other inspection as part of a PMI. Who is correct?

    A. A only
    B. B only
    C. Both A and B
    D. Neither A nor B

40. A vehicle with vacuum boosted brakes is tagged out of service if in the vacuum system the:

    A. Vacuum never exceeds 15 in. Hg.
    B. Vacuum lines appear to be OK, but the engine lacks performance.
    C. Vacuum reserve is insufficient to permit one full brake application after the engine is shut off.
    D. System vacuum bleeds down over 50 percent after engine shut off within two minutes.

41. Technician A says that slack adjusters should be lubricated when brakes are inspected. Technician B says that only the rear brake chamber pushrods have to be checked for stress cracks. Who is correct?

    A. A only
    B. B only
    C. Both A and B
    D. Neither A nor B

42. When adjusting wheel bearings, Technician A says that TMC wheel-end procedure should be followed at all times. Technician B says that wheel bearings set with preload are likely to fail prematurely. Who is correct?

    A. A only
    B. B only
    C. Both A and B
    D. Neither A nor B

43. Technician A says any type of wire may be used to replace a burned fusible link as long as the gauge is one size smaller than the wire in the circuit being protected. Technician B says that a circuit breaker should be replaced after repairing a short in a circuit. Who is correct?

    A. A only
    B. B only
    C. Both A and B
    D. Neither A nor B

44. After a stretch of rough road is driven, the steering wheel continues to shake for a few seconds. Which of these could be the cause?

    A. Leaking front shock absorbers
    B. Low tire pressure
    C. Rusted rear shock absorbers
    D. Missing jounce bumper

45. Which of the following is the standard fifth wheel plate height?

    A. 39 inches
    B. 43 inches
    C. 47 inches
    D. 49 inches

46. A maintenance-free battery is low on electrolyte and the built-in hydrometer indicates yellow. Technician A says a faulty voltage regulator may cause this problem. Technician B says high resistance in the circuit between the alternator battery terminal and the positive battery terminal may be the cause. Who is correct?

    A. A only
    B. B only
    C. Both A and B
    D. Neither A nor B

47. On an automatic transmission, the following methods can be used for checking transmission shift points EXCEPT:

    A. Stall test method.
    B. Speedometer readings.
    C. Test stand tailshaft speed.
    D. Engine RPM.

48. A highway tractor is having a full PMI performed. Technician A says that outboard S-cam bushings should be pressure-lubricated until grease flows from the outboard seal. Technician B says to use the maximum pressure available from pressure lubricating equipment to aid in flushing contaminants from lubricated components. Who is correct?

    A. A only
    B. B only
    C. Both A and B
    D. Neither A nor B

49. An excessively high coolant level in the recovery reservoir may be caused by any of these problems EXCEPT:

    A. Restricted radiator tubes.
    B. A thermostat that is stuck open.
    C. A blown head gasket.
    D. An inoperative air-operated cooling fan clutch.

50. Wheel assemblies are being discussed. Technician A says hub-piloted disc wheels are less susceptible to loosening and cracking than stud-piloted wheels. Technician B says that disc wheels generally experience more alignment and balance problems than do cast spoke designs. Who is correct?

    A. A only
    B. B only
    C. Both A and B
    D. Neither A nor B

PREPARATION EXAM 3

1. Technician A says that all current tractor-trailers are required to have automatic slack adjusters. Technician B says that automatic slack adjusters increase the legal free stroke dimension. Who is correct?

   A. A only
   B. B only
   C. Both A and B
   D. Neither A nor B

2. An alternator with a 90-ampere rating produces 45 amperes during an output test. What is the most likely cause of this problem?

   A. An alternator drive belt that is bottomed in the pulley
   B. An open circuit in the rotor field coil
   C. A misaligned alternator pulley
   D. A broken alternator brush lead

3. All of these conditions could be Commercial Vehicle Safety Alliance (CVSA) out-of-service criteria EXCEPT:

   A. An exhaust leak under the sleeper.
   B. A wiring harness that contacts the exhaust.
   C. An inaccurate exhaust temperature gauge.
   D. A missing exhaust heat shield.

4. A slack adjuster is being adjusted. Technician A says wear between the clevis pin and yoke will cause the stroke to be too long. Technician B says that an automatic slack adjuster will need frequent adjustment. Who is correct?

   A. A only
   B. B only
   C. Both A and B
   D. Neither A nor B

5. What should you do to eliminate a mildew smell from the air conditioner (A/C) output air?

   A. Remove the evaporator case, clean it with a vinegar and water solution, and dry it thoroughly before reinstallation.
   B. Spray a disinfectant into the panel outlets.
   C. Pour a small amount of alcohol into the air intake plenum.
   D. Place an automotive deodorizer under the dash.

6. An inversion valve is being tested on a heavy-duty truck. Technician A says to drain both the supply tank and the primary circuit reservoir before beginning the test. Technician B verifies that secondary circuit pressure is at least 100 psi before conducting the test. Who is correct?

    A. A only
    B. B only
    C. Both A and B
    D. Neither A nor B

2013 © Delmar, Cengage Learning

7. What type of battery do the arrows in the illustration above indicate?

    A. Top terminal, low maintenance
    B. Stud terminal, maintenance free
    C. Side terminal, maintenance free
    D. Inverted terminal, high maintenance

8. Technician A says that a possible cause of an above-normal engine temperature is a faulty fan clutch. Technician B claims that plugged radiator fins may cause overheating. Who is correct?

    A. A only
    B. B only
    C. Both A and B
    D. Neither A nor B

9. If the floating caliper on a hydraulic brake truck is seized, which of the following conditions will result?

    A. Brake pedal fade
    B. Brake pedal pulsations
    C. Reduced braking force
    D. Brake grabbing

10. A windshield is being inspected during a routine inspection. Technician A says that the vehicle need not be deadlined, since there are no cracks in the windshield greater than 2 inches in length on the driver's side of the vehicle. Technician B says that any vision distorting defect in the windshield glass must be within the sweep of the wiper on the driver's side to deadline the vehicle. Who is correct?

    A. A only
    B. B only
    C. Both A and B
    D. Neither A nor B

11. A technician notices a whitish milky substance when changing the fluid in an axle. What is the most likely cause of this evidence?

    A. Common condensation
    B. Infrequent driving and short trips
    C. The axle being submerged in water
    D. The vehicle being frequently driven during rainy or wet conditions

12. Technician A says that VORAD stands for variable output relay accessory device. Technician B says that anti-collision devices are now mandatory on all heavy-duty vehicles. Who is correct?

    A. A only
    B. B only
    C. Both A and B
    D. Neither A nor B

13. When checking headlight aiming, Technician A says that the vehicle should be on level ground at a distance of 35 feet from a screen. Technician B says you must always use headlight aiming equipment when adjusting headlights. Who is correct?

    A. A only
    B. B only
    C. Both A and B
    D. Neither A nor B

14. The result of a shorted field winding inside an alternator rotor is:

    A. No magnetic strength around the rotor.
    B. Low current flow through the field winding.
    C. Excessive alternator voltage output.
    D. Damage to the voltage regulator.

15. Technician A checks coolant hoses for cracks and hardness every time a general inspection is made. Technician B says that a general coolant system inspection should be made every time a vehicle is scheduled for a PMI. Who is correct?

    A. A only
    B. B only
    C. Both A and B
    D. Neither A nor B

16. When performing a hydrometer test on a truck battery, all of the following are incorrect EXCEPT:

    A. If the battery temperature is 50°F (10°C), the battery should be brought to room temperature before testing.

    B. If the battery temperature is 120°F (49°C), 0.020 specific gravity points should be subtracted from the hydrometer reading.

    C. The maximum variation in cell hydrometer readings is 0.100 specific gravity points.

    D. The battery is fully charged if all the cell hydrometer readings are above 1.265.

17. When discussing a battery capacity test with the battery at 70°F (21°C), Technician A says the battery discharge rate is two times the battery reserve capacity rating. Technician B says the battery is satisfactory if the voltage remains above 9.2 V. Who is correct?

    A. A only

    B. B only

    C. Both A and B

    D. Neither A nor B

18. A viscous vibration damper is being inspected. What would be the most serious consequence of a visible dent in the housing?

    A. Engine crankshaft failure

    B. Engine misfire

    C. Engine surging

    D. Knocking noise

19. Technician A says to inflate mounted tires in a safety cage or using a portable lock ring guard. Technicians B says to first mount the tire on the truck and then inflate to the proper tire pressure. Who is correct?

    A. A only

    B. B only

    C. Both A and B

    D. Neither A nor B

20. After removing a 30-ampere circuit breaker from a fuse panel, a technician tests breaker continuity with an ohmmeter. Technician A says the current from the ohmmeter will open the circuit breaker. Technician B says the ohmmeter should read infinity if the circuit breaker is satisfactory. Who is correct?

    A. A only

    B. B only

    C. Both A and B

    D. Neither A nor B

2013 © Delmar, Cengage Learning

21. An audible air leak is observed that appears to be coming from the front service brake chamber in the illustration above during an annual inspection. Which of the following is the LEAST LIKELY cause?

A. Loose mounting bolts

B. Ruptured diaphragm

C. Loose chamber clamp

D. Missing clamp bolt

22. The current flow in an electric circuit is lower than specified. Technician A says the voltage supplied to the circuit may be higher than specified. Technician B says the resistance in the circuit may be higher than specified. Who is correct?

A. A only

B. B only

C. Both A and B

D. Neither A nor B

23. While discussing van and over-top trailer bodies, Technician A says if one floor cross member is broken and sagging below the lower rail, the vehicle should be tagged out of service. Technician B says if a drop frame is visibly twisted, the vehicle should be placed out of service. Who is correct?

A. A only

B. B only

C. Both A and B

D. Neither A nor B

24. While inspecting a transmission for leaks, the technician notices that the gaskets appear to be blown out of their mating sealing surfaces. What should the technician check first?

A. The transmission breather

B. The shifter cover

C. The release bearing

D. The rear seal

25. A digital electronic service tool (EST) is connected to a truck databus during a routine PMI to read engine data. Technician A says active trouble codes cannot be erased. Technician B says that an event audit trail can be read. Who is correct?

    A. A only
    B. B only
    C. Both A and B
    D. Neither A nor B

26. Technician A says any type of wire may be used to replace a burned fusible link as long as the gauge is one size smaller than the wire in the circuit being protected. Technician B says that a circuit breaker should be replaced after repairing a short in a circuit. Who is correct?

    A. A only
    B. B only
    C. Both A and B
    D. Neither A nor B

2013 © Delmar, Cengage Learning

27. What is the item indicated by the arrow in the illustration above?

    A. A water pump pulley
    B. An air pump pulley
    C. A fuel pump pulley
    D. An idler pulley

28. Technician A says that air drum brake linings should be replaced when there are less than ¼ inch of linings when measured at the center of the shoe. Technician B says that brake linings should be replaced when they are saturated with oil or grease. Who is correct?

    A. A only
    B. B only
    C. Both A and B
    D. Neither A nor B

29. Which of the following cause most heavy-duty pneumatic fan clutch failures?

    A. Seized support bearings
    B. Air leaks
    C. Electrical shorts
    D. Obstructions

30. Technician A says to block the wheels before adjusting the service brakes on a truck. Technician B says that the air reservoir pressure must be dropped below the cut-in value before the air compressor will resume its effective cycle. Who is correct?

    A. A only
    B. B only
    C. Both A and B
    D. Neither A nor B

31. During a PMI, the technician tests the air supply system and finds that buildup time is slow. Which of the following is the most likely cause?

    A. A defective air governor
    B. A leak in a service brake chamber
    C. An air leak in the cab
    D. A defective air compressor

32. The trailer lights of a double combination vehicle are being inspected. Technician A says that the wiring connector for all trailer lights is usually an 8-pin connector. Technician B says that it is mandatory to check the lights on the last trailer first in a multiple combination vehicle. Who is correct?

    A. A only
    B. B only
    C. Both A and B
    D. Neither A nor B

33. A vehicle is being inspected for a safety certification. Technician A states if wetness is seen around a seam of a fuel tank without forming any drops, it is passable. Technician B says that it is not necessary to drain water from the fuel when performing an inspection during the summer months. Who is correct?

    A. A only
    B. B only
    C. Both A and B
    D. Neither A nor B

2013 © Delmar, Cengage Learning

34. Which of these conditions do the shoes in the illustration above show?

   A.   Unequal wear
   B.   Poor contact at center of linings
   C.   Lining material tapered
   D.   Normal wear

35. Technician A says overfilling a manual transmission could result in overheating of the transmission. Technician B says overfilling a transmission could cause excessive aeration of the transmission fluid. Who is correct?

   A.   A only
   B.   B only
   C.   Both A and B
   D.   Neither A nor B

36. When checking for diesel engine fuel system leaks, the engine must be:

   A.   Cold.
   B.   At normal operating temperature.
   C.   Running.
   D.   Off.

37. Technician A says that an inspection of the road speed governor is required during the pre-trip inspection. Technician B says that more emphasis should be placed on the post-trip inspection rather than the pre-trip inspection. Who is correct?

   A.   A only
   B.   B only
   C.   Both A and B
   D.   Neither A nor B

38. Fluid leaks are evident between the axle housing and the carrier assembly. What would the most likely cause of this leak be?

   A.   Damaged gasket or missing sealant
   B.   Repeated overloading of the drive train
   C.   Plugged axle housing breather vent
   D.   Moisture contaminated axle lubricant

2013 © Delmar, Cengage Learning

39. Technician A says that the dash control button on the left in the illustration above is the trailer service brake valve. Technician B says that the dash control button on the right in the same illustration is the system park valve. Who is correct?

    A. A only

    B. B only

    C. Both A and B

    D. Neither A nor B

40. What circuit is being operated when a starter draw test is being performed?

    A. Charging circuit

    B. Cranking circuit

    C. Regulator circuit

    D. Control circuit

2013 © Delmar, Cengage Learning

41. The bearings in the wheel hub shown in the illustration above:

    A. Must be lubricated with wheel bearing grease before assembly.

    B. Are permanently lubricated with the specified oil.

    C. Require a specific hub end-play.

    D. Require inner and outer hub seals.

42. During a PMI, the antilock brake system (ABS) warning light is illuminated with the engine running. Technician A says there is a mechanical defect in the ABS. Technician B says the ABS will still function normally. Who is correct?

   A. A only
   B. B only
   C. Both A and B
   D. Neither A nor B

2013 © Delmar, Cengage Learning

43. Technician A says the device in the illustration above is to restrain the tire while it is being inflated. Technician B says that the device could be used to center the tire for reuse. Who is correct?

   A. A only
   B. B only
   C. Both A and B
   D. Neither A nor B

44. What can result from a shock absorber outer tube touching the fluid reservoir?

   A. Premature bushing wear
   B. Scraping noise with travel
   C. Main shaft seal failure
   D. Excessive fatigue in mount assembly

45. Technician A says that when refilling a cooling system it may be necessary to open a bleed valve. Technician B says that when checking a cooling system that uses ELC, to check the coolant conditioner level by using an SCA test strip specifically designed for ELC coolants. Who is correct?

   A. A only
   B. B only
   C. Both A and B
   D. Neither A nor B

46. A scheduled inspection is being performed. Technician A says that if lift gates and controls are not on the checklist, then it is not mandatory to inspect them. Technician B says you always check cargo ramps on a vehicle equipped with them. Who is correct?

   A. A only
   B. B only
   C. Both A and B
   D. Neither A nor B

47. The vehicle to which a pintle hook or drawbar is attached should be tagged out of service for any of the these defects EXCEPT:

   A. Wear in the pintle hook horn that exceeds 10 percent of the horn section.
   B. Loose or missing mounting bolts.
   C. Any welded repairs to the drawbar eye.
   D. Cracks in the pintle hook mounting surface.

48. A truck's clutch is being measured for its free travel. Technician A says the measurement should be between 1 and 2 inches. Technician B says that on a hydraulic clutch, there should be no free travel. Who is correct?

   A. A only
   B. B only
   C. Both A and B
   D. Neither A nor B

49. What would be the LEAST LIKELY cause of a front wheel shimmy?

   A. Kingpin play
   B. Improperly adjusted steering gear relief plungers
   C. Loose Pitman arm
   D. Tire imbalance

50. Technician A says that a proper inspection of the cooling system should include a check of coolant-usage records. Technician B changes the coolant at every PMI. Who is correct?

   A. A only
   B. B only
   C. Both A and B
   D. Neither A nor B

PREPARATION EXAM 4

1. A truck mechanically actuated clutch is being measured for freeplay. Technician A says the measurement should be between 1 and 2 inches. Technician B says that on a hydraulic clutch freeplay should be 2 to 3 inches. Who is correct?

   A. A only
   B. B only
   C. Both A and B
   D. Neither A nor B

2. Which of the following defines the cloud point of diesel fuel?

   A. The lowest temperature at which it will flow
   B. The temperature at which a haze of wax crystals will form
   C. The highest temperature before evaporation begins
   D. The temperature at which the fuel produces black smoke

3. Technician A says you turn the key on the ignition to perform electronic system self-checks. Technician B says the ignition circuit powers most computer-controlled chassis systems. Who is correct?

    A. A only

    B. B only

    C. Both A and B

    D. Neither A nor B

2013 © Delmar, Cengage Learning

4. What type of connections are pictured in the illustration above?

    A. Parallel

    B. Parallel-series

    C. Series

    D. Series-parallel

5. When performing a test drive on a vehicle, Technician A says to test the service brakes by applying and releasing them. Technician B says that although a test drive is not mandatory, it is a good idea to perform one after a vehicle has been serviced. Who is correct?

    A. A only

    B. B only

    C. Both A and B

    D. Neither A nor B

6. In addition to replacing the fuel filter and priming the fuel system, what else should be performed when doing a preventive maintenance inspection (PMI) on a fuel system?

    A. Check the fuel tank mountings.

    B. Test fuel pour point.

    C. Check heater lines for proximity to the fuel lines.

    D. Recalibrate the fuel gauge.

7. An inspection is being performed on a heavy-duty truck brake system and grease soaked linings are found on three wheels of the trailer's rear tandem. Technician A says to inform the driver. Technician B writes up the problems and informs the driver. Who is correct?

    A. A only
    B. B only
    C. Both A and B
    D. Neither A nor B

8. When winterizing a truck, which of the following procedures should be done first?

    A. Replacing lost coolant
    B. Cleaning the coolant
    C. Adding antifreeze
    D. Testing the condition of the coolant

9. Which of these criteria is the LEAST LIKELY to be considered tire mismatching?

    A. Radials and bias ply on the same axle
    B. Dual wheels of the same nominal size but different tread design
    C. Installation of a new tire next to a worn-out tire on the same axle
    D. Tires with different circumferences on the same axle

10. Technician A says the vibration damper counterbalances the back-and-forth twisting motion of the crankshaft each time a cylinder fires. Technician B says if the seal contact area on the vibration damper hub is scored, the damper assembly must be replaced. Who is correct?

    A. A only
    B. B only
    C. Both A and B
    D. Neither A nor B

2013 © Delmar, Cengage Learning

11. What is being performed in the illustration above?

    A. Lateral runout and rotor thickness
    B. Wheel bearing runout
    C. Rotor thickness and radial runout
    D. Parallelism

12. A starting motor fails to crank the engine and indicates a very high current draw. The most likely cause of this problem is:

    A. A direct ground in the starter field coils.

    B. An open circuit in the armature windings.

    C. Broken brush springs and no tension on the brushes.

    D. Failed magnetic switch.

13. A fleet manager is restructuring a PMI program. Technician A says that fire extinguisher distribution varies with the type of vehicle it is installed in. Technician B says that it is the fleet manager's responsibility to see that personnel performing annual inspections are qualified. Who is correct?

    A. A only

    B. B only

    C. Both A and B

    D. Neither A nor B

14. An air brake system has slower than specified air pressure buildup time. What could be the cause of this problem?

    A. A smaller than specified supply reservoir

    B. An air leak in the compressor air intake filter

    C. Carbon buildup in the compressor discharge hose

    D. An excessive amount of water in the supply reservoir

15. Visual inspection of the power steering belt(s) can reveal all the following EXCEPT:

    A. Proper belt tension.

    B. Premature wear due to misalignment.

    C. Correct orientation of dual belt application.

    D. Proper belt seating in the pulley.

16. A driver complains of a door not closing properly and has brought it in for service. Technician A immediately greases the hinges. Technician B retrieves the vehicle records and checks for past problems with the doors and makes a note of the driver reports. Who is correct?

    A. A only

    B. B only

    C. Both A and B

    D. Neither A nor B

17. Technician A always begins battery inspection with a general visual first, followed by an open circuit voltage test. Technician B says that it may be necessary to remove the surface charge from a battery before it is open circuit voltage tested. Who is correct?

    A. A only

    B. B only

    C. Both A and B

    D. Neither A nor B

18. A slack adjuster pushrod is being serviced on a vehicle following a driver tag on a post-trip inspection. Technician A says that the service performed must be logged before the vehicle can be put back into service. Technician B says the service performed should be recorded on a "B" PMI maintenance sheet. Who is correct?

    A. A only

    B. B only

    C. Both A and B

    D. Neither A nor B

19. What is being performed in the illustration above?

    A. Testing the battery output

    B. Checking the continuity of the ignition switch

    C. Performing a voltage drop on the ground circuit side of the starter

    D. By-passing the magnetic switch

20. Technician A says an open circuit voltage test may be used in place of a hydrometer test. Technician B says during the open circuit voltage test the headlights should be on. Who is correct?

    A. A only

    B. B only

    C. Both A and B

    D. Neither A nor B

21. In relation to lubricants (transmission and drive axle), what is channeling?

    A. Lubricant breakdown in high temperatures

    B. Extreme pressures on components causes a lubricant to separate

    C. Liquid seeking the lowest center of gravity

    D. Components starved of lubrication due to cold temperatures

22. Technician A says that defective and/or improper ground connections are the most common sources of dim lights. Technician B says that when vehicle lighting inspections are performed, it is common practice to start at the back of the trailer and move forward. Who is correct?

    A. A only
    B. B only
    C. Both A and B
    D. Neither A nor B

23. All of the following would cause a constant leak from the air dryer purge valve EXCEPT:

    A. Foreign particles on valve seat.
    B. A purge valve seal damaged.
    C. A purge valve frozen.
    D. A restricted purge line.

24. Technician A says that S-cam-actuated foundation brakes are the most common brake system on heavy-duty trucks today. Technician B says that slack adjusters and S-cams convert linear motion into brake torque. Who is correct?

    A. A only
    B. B only
    C. Both A and B
    D. Neither A nor B

25. The diameter on a bake drum is greater at the edges of the friction surfaces than in the center. This drum condition is:

    A. Bell-mouthed drum.
    B. Concave brake drum.
    C. Drum out-of-round.
    D. Convex brake drum.

26. A digital diagnostic reader (DDR) is connected to the American Trucking Association (ATA) connector during a routine PMI to read engine data. Technician A says active codes can be read. Technician B says that historic codes can be erased. Who is correct?

    A. A only
    B. B only
    C. Both A and B
    D. Neither A nor B

2013 © Delmar, Cengage Learning

27. What does the arrow in the illustration above indicate?

    A.   Clearance light

    B.   Identification light

    C.   Fog light

    D.   Intermediate light

28. While performing routine maintenance, a technician observes that the power steering system
    has a filter. Technician A says the filter should be replaced, if required, on a preventive
    maintenance inspection (PMI) according to the OEM. Technician B says the filter should be
    replaced after a component failure. Who is correct?

    A.   A only

    B.   B only

    C.   Both A and B

    D.   Neither A nor B

29. All of the following could cause the power-steering-pump pulley to become
    misaligned EXCEPT:

    A.   An over-pressed pulley.

    B.   Loose fit from pulley hub to pump shaft.

    C.   A loose pump mounting bracket.

    D.   A broken engine mount.

30. Technician A says that air horns should be checked during a PMI. Technician B says any
    physical damage found on the front bumper of a trailer should be noted during a PMI. Who
    is correct?

    A.   A only

    B.   B only

    C.   Both A and B

    D.   Neither A nor B

31. Technician A says that a minor cooling system leak is sufficient criteria for deadlining. Technician B says that an engine should be checked for both static and dynamic leaks. Who is correct?

    A. A only
    B. B only
    C. Both A and B
    D. Neither A nor B

2013 © Delmar, Cengage Learning

32. A fifth wheel is being mounted on a tractor. Technician A says that if it is located too far forward, hard steering and tire wear can result. Technician B says the fifth wheel should be mounted to the frame flanges by welding. Who is correct?

    A. A only
    B. B only
    C. Both A and B
    D. Neither A nor B

33. With the brakes applied at 80 psi (551 kPa), and checking for maximum allowable leakage at the brake application valve exhaust port, all of the following are incorrect EXCEPT:

    A. A 2-inch (50.8 mm) bubble in 1 second.
    B. A 2-inch (50.8 mm) bubble in 2 seconds.
    C. A 3-inch (76.2 mm) bubble in 3 seconds.
    D. A 1-inch (25.4 mm) bubble in 3 seconds.

34. When diagnosing an ABS with an ABS check switch and a light in the instrument panel, all of the following are incorrect EXCEPT:

    A. When the ABS check switch and the ignition switch are turned on, the ABS light flashes blink codes.
    B. The first set of flashes in the blink code sequence indicates whether the system is ABS only or ABS and automatic traction control (ATC).
    C. When there are no electrical defects in the ABS, a 0-0-0 blink code is revealed.
    D. A blink code indicates whether an electrical defect is in a wheel speed sensor or the connecting wiring.

35. A customer has a truck towed into the shop and says the starter would not crank the engine over. What should be checked first?

    A. The ground cable connection
    B. The starter solenoid circuit
    C. The ignition switch crank circuit
    D. The battery condition

36. Technician A says that an engine can be checked for leaks while the engine is off. Technician B says to also check the engine for leaks while it is running. Who is correct?

    A. A only
    B. B only
    C. Both A and B
    D. Neither A nor B

37. All of the following could cause excessive leakage in an air brake system with the service brakes applied EXCEPT:

    A. A leaking brake chamber diaphragm.
    B. A leaking hose, tube, or fitting.
    C. Defective compressor gaskets.
    D. A bad relay valve.

38. Technician A says a correctly adjusted pull type clutch requires ½ inch (1.25 cm) of clearance between the release bearing and the clutch brake disc. Technician B says a correctly adjusted pull type clutch needs ⅛ inch (0.3125 cm) clearance between itself and the release bearing. Who is correct?

    A. A only
    B. B only
    C. Both A and B
    D. Neither A nor B

39. When inspecting driveshafts, all of the following can result in vibration EXCEPT:

    A. Missing balance weight.
    B. Out of phase condition.
    C. Light rust on the driveshaft surface.
    D. Loose U-joint cap.

40. Technician A says a battery drain of 2 amperes would cause the battery to become discharged. Technician B says the battery drain test should be performed with the engine at normal operating temperature. Who is correct?

    A. A only
    B. B only
    C. Both A and B
    D. Neither A nor B

41. During an inspection, a tractor must be placed out of service if any of these tire conditions are present EXCEPT:

    A. If a tire is mounted or inflated so it contacts any part of the vehicle.

    B. If a drive axle tire has a tread depth of less than 1/32 inch at two adjacent tread grooves at three separate locations.

    C. If a steer axle tire has a tread depth of 5/32 inch at two adjacent tread grooves at any location on the tire.

    D. If the tire is labeled "Not for Highway Use".

42. An electronic cruise control maintains the vehicle speed at 45 mph (72 km/h) when the selected speed is 60 mph (96 km/h). What is the most likely cause of this problem?

    A. A defective cruise control module

    B. A defective vehicle speed sensor (VSS)

    C. Improper size of tires on the drive axles

    D. A defective cruise control switch

43. During a tire inspection, a technician notices a cut in the sidewall of a tire. Which of the following is the most likely cause?

    A. Excessive toe-in

    B. Camber

    C. Careless driving

    D. Caster

44. Technician A says that impact guards are required to be equipped on vehicles manufactured after 1998. Technician B says that the impact guards should be carefully inspected for damage during routine PMI inspection since they are subject to abuse. Who is correct?

    A. A only

    B. B only

    C. Both A and B

    D. Neither A nor B

45. A turbocharger is being inspected for oil leaks. Technician A says to check it while the engine is running. Technician B measures turbine shaft radial runout with the engine running. Who is correct?

    A. A only

    B. B only

    C. Both A and B

    D. Neither A nor B

46. All of the following are underside components of a trailer EXCEPT:

    A. Bulkhead.

    B. Tires.

    C. Suspension.

    D. Landing gear.

47. An exhaust brake is being inspected on a truck. Technician A says that this should be performed during the road test. Technician B says that a properly functioning engine compression brake should be noiseless. Who is correct?

    A. A only
    B. B only
    C. Both A and B
    D. Neither A nor B

48. A cooling system on a heavy truck is being tested. Technician A says that the color of the coolant indicates whether it is EG, PG, or ELC. Technician B says to drain off a gallon of coolant before performing a pressure test. Who is correct?

    A. A only
    B. B only
    C. Both A and B
    D. Neither A nor B

49. Which of the following would be the least important part of a trailer's lighting system to check?

    A. Side amber marker light(s)
    B. Interior overhead light(s)
    C. Main wiring input socket and associated wiring
    D. Reefer power supply

50. Technician A says that retorquing wheel lug nuts are not required when performing a PMI. Technician B says that increased load carrying capacity is an advantage of switching to low-profile tires. Who is correct?

    A. A only
    B. B only
    C. Both A and B
    D. Neither A nor B

PREPARATION EXAM 5

1. Technician A says that a fully charged battery will not freeze as easily as a discharged battery. Technician B says that the specific charge of a battery has no effect on the electrolyte freezing. Who is correct?

    A. A only
    B. B only
    C. Both A and B
    D. Neither A nor B

2. A truck is being road tested as part of a scheduled PMI. All of the following should be checked EXCEPT:

   A. Check road speed limiter.

   B. Perform snap-shot test for active codes.

   C. Observe exhaust for excessive smoke.

   D. Verify engine brake operation.

3. Technician A says that lubricant leaking from the drive axle seals could be caused by a plugged breather. Technician B says that grease fittings should be wiped before lubricating. Who is correct?

   A. A only

   B. B only

   C. Both A and B

   D. Neither A nor B

4. When checking the coolant condition with a supplemental coolant additive (SCA) test strip, the technician finds that the coolant pH condition is higher than specification limits. While reviewing maintenance records the technician finds that the pH condition was higher than the limit at the last service also. Which of these items should the technician do?

   A. Add more antifreeze to increase SCA.

   B. Continue to run the truck until the next PMI.

   C. Drain the entire cooling system and add the proper SCA mixture.

   D. Run the truck with no SCA additives until the next PMI.

2013 © Delmar, Cengage Learning

5. Which of the following would not be part of performing a PMI on a truck with a fifth wheel, like the one shown in the illustration above?

   A. Reverse the location of the jaws every 100,000 miles.

   B. Make sure horizontal movement does not exceed ⅛ inch.

   C. Torque mounting bolts.

   D. Inspect the locking mechanism.

6. What would be the LEAST LIKELY cause of a high-amperage reading when checking starter current draw?

   A. Short in the starter motor

   B. Mechanical resistance due to binding engine

   C. Starter wire gauge is too small

   D. Short to ground in the starter wire

7. Technician A says that all electronic ABS are equipped with self-diagnostics. Technician B says that ABS is an option on new trucks. Who is correct?

    A. A only
    B. B only
    C. Both A and B
    D. Neither A nor B

8. Technician A says that a fuel leak while the engine is running is less hazardous than when the engine is off. Technician B says that an engine should be checked for both static and dynamic leaks. Who is correct?

    A. A only
    B. B only
    C. Both A and B
    D. Neither A nor B

9. Which of the following is the LEAST LIKELY cause of water droplets or gray oil on the dipstick?

    A. Cracked block or cylinder head
    B. Blown head gasket
    C. Leaky oil cooler
    D. Worn oil pump

10. The major changes introduced by the Clean Air Act in heavy-duty truck air conditioning (A/C) systems relate to:

    A. Certification of A/C technicians.
    B. Maintaining older equipment.
    C. Replacing leaking equipment.
    D. Refrigerants and their handling.

11. Technician A says that the tractor protection valve protects the tractor air supply under trailer breakaway conditions. Technician B says that the trailer supply dash valve (emergency) is diamond shaped and yellow in color. Who is correct?

    A. A only
    B. B only
    C. Both A and B
    D. Neither A nor B

12. A small leak is found in an A/C system. What should the technician do?

    A. Nothing. Wait for a service report to be filed.
    B. Log the problem on a PMI schedule sheet.
    C. Log the problem, and then inform the owner/driver that this is in violation of the Clean Air Act.
    D. Repair the leak immediately.

13. Technician A says the Department of Transportation (DOT) does not require PMI records on lubrication. Technician B says that maintenance records should be referenced before undertaking major repairs. Who is correct?

   A. A only
   B. B only
   C. Both A and B
   D. Neither A nor B

14. Technician A says that the link between the brake chamber and the foundation brake is known as the brake shoe roller. Technician B says that the link between the brake chamber and the foundation brake is the slack adjuster. Who is correct?

   A. A only
   B. B only
   C. Both A and B
   D. Neither A nor B

15. Technician A states that a vehicle should be deadlined if weld repairs are found to have been made on pintle chains and hooks. Technician B says that at least one functioning amber or red light on rear loads projecting more than 4 feet beyond the vehicle body is required. Who is correct?

   A. A only
   B. B only
   C. Both A and B
   D. Neither A nor B

2013 © Delmar, Cengage Learning

16. In the illustration above, what could commonly be mounted above the roof depicted by arrow?

   A. Identification lights
   B. Solar panels
   C. Skylight
   D. Air vents for sleeper cab

17. Technician A says that fuel tanks should be left near empty after long runs before refilling. Technician B says that after a long run, the fuel tanks should be refilled and not left empty. Who is correct?

    A. A only
    B. B only
    C. Both A and B
    D. Neither A nor B

18. Technician A says that turbochargers perform at very high speeds which can be in excess of 100,000 rpm. Technician B says that turbochargers only turn at engine RPM. Who is correct?

    A. A only
    B. B only
    C. Both A and B
    D. Neither A nor B

19. Technician A says that the condition of the batteries should be verified with a load test prior to performing an alternator current output test. Technician B says that a load tester with an inductive pick-up may be used to test alternator current output. Who is correct?

    A. A only
    B. B only
    C. Both A and B
    D. Neither A not B

20. Technician A says that when checking brake chamber stroke using the applied stroke method to use 25 to 35 psi. Technician B says that when checking brake chamber stroke using the applied stroke method to use only 10 psi. Who is correct?

    A. A only
    B. B only
    C. Both A and B
    D. Neither A nor B

21. What is the approximate stoichiometric air-fuel ratio?

    A. 15:1
    B. 8:1
    C. 10:1
    D. 17.5:1

22. Technician A says that a driver must receive a visible alert when system air pressure drops below 100 psi. Technician B says that to check for proper functioning of the inversion (spring brake valve) valve, the air system should be at 60 psi. Who is correct?

    A. A only
    B. B only
    C. Both A and B
    D. Neither A nor B

23. The driver complains about a rotten egg smell in the truck cab that is most noticeable when the A/C system is first turned on. What could be the cause of this problem?

    A. Leaking evaporator core

    B. Leaking condenser core

    C. Leaking heater core

    D. Plugged evaporator case drain

2013 © Delmar, Cengage Learning

24. What is the procedure being performed in the illustration above?

    A. Knuckle vertical play

    B. Knuckle upper bushing play measurement

    C. Lower bushing deflection

    D. Kingpin axial play

25. All of these procedures should be done when replacing a torque rod on a tandem axle suspension EXCEPT:

    A. Replace alignment washers with new ones.

    B. Secure the truck's parking brakes and chock the tires.

    C. Ensure the replacement torque rod is of the correct length.

    D. Check the axle alignment when finished.

26. Technician A says some manufacturers recommend the secondary fuel filter be installed dry and primed with a hand pump. Technician B says that most fuel filters are changed at routine intervals (preventive maintenance inspection schedules). Who is correct?

    A. A only

    B. B only

    C. Both A and B

    D. Neither A nor B

27. Which of the following is the best description of an acceptable axle lubricant level?

    A. Lube level is even with the bottom of the filler hole

    B. Lube level can be easily seen through the filler hole

    C. Lube can be touched one knuckle joint below the bottom of the filler hole

    D. Lube can be touched with the index finger through the filler hole

28. Technician A says it is good practice to check underground fuel storage tanks for water content. Technician B says batteries can be kept from freezing if they have a full charge. Who is correct?

   A. A only
   B. B only
   C. Both A and B
   D. Neither A nor B

29. A worn front steering bushing can cause all of the following EXCEPT:

   A. A loose feeling in the steering wheel when driving straight.
   B. Noise when hitting a bump in the road.
   C. Excessive use of power steering fluid.
   D. Pull to the left while driving.

30. Technician A says that if the temperature of a wire increases, its resistance will decrease. Technician B says that if the temperature of a wire increases, its resistance will increase. Who is correct?

   A. A only
   B. B only
   C. Both A and B
   D. Neither A nor B

31. What is the most common cause of brake imbalance on a truck or trailer?

   A. Different size brake chambers
   B. Brakes out of adjustment
   C. Mismatched brake linings
   D. Mismatched brake hose sizes

32. A truck has a slower than normal air pressure buildup time. Technician A says that leaking air lines, connections, or valves could be the cause. Technician B says a worn air compressor could be the cause. Who is correct?

   A. A only
   B. B only
   C. Both A and B
   D. Neither A nor B

33. Technician A says a circuit breaker is designed to protect a circuit from high current. Technician B says a circuit breaker's rating is given in amps. Who is correct?

   A. A only
   B. B only
   C. Both A and B
   D. Neither A nor B

34. Which of the following should be done when performing a preventive maintenance inspection (PMI) on a heavy truck fuel system?

    A. Check the pour point of the fuel.
    B. Verify the accuracy of the fuel gauge.
    C. Check the fuel pump for volumetric capacity.
    D. Replace the fuel filter and prime the system.

35. Technician A says that a circuit should be energized when using an ohmmeter. Technician B says that a voltage drop test is an accurate means of identifying high resistance in an energized circuit. Who is correct?

    A. A only
    B. B only
    C. Both A and B
    D. Neither A nor B

36. Technician A says that the primary tank feeds the supply tanks in an air brake system. Technician B says that the supply tank feeds the primary and secondary tanks in an air brake system. Who is correct?

    A. A only
    B. B only
    C. Both A and B
    D. Neither A nor B

37. Technician A says that an open circuit can be caused by an open switch. Technician B says that an open circuit can be caused by a broken wire. Who is correct?

    A. A only
    B. B only
    C. Both A and B
    D. Neither A nor B

38. What is the best way to clean a rear-axle breather?

    A. Replace the breather.
    B. Soak the breather in gasoline.
    C. Soak in solvent, and then blow out with compressed air.
    D. Use a rag to wipe the vent orifice clean.

39. Technician A says that when considering whether to deadline a vehicle, it is necessary to weigh the importance of getting a load moved versus a mechanical problem that could cause a breakdown but is not safety related. Technician B says that a vehicle should be deadlined for a cruise control malfunction. Who is correct?

    A. A only
    B. B only
    C. Both A and B
    D. Neither A nor B

40. What should a technician do to thoroughly inspect a transmission mount?

    A. Remove the mount and tension the two mounting plates.
    B. Remove the mount and inspect for signs of damage.
    C. Visually inspect the mount while still in the vehicle.
    D. Replace the mount if it is suspect.

41. Technician A says that steer axle tires must have at least 2/32 inch remaining at the most worn spot. Technician B says that tires on the drive or trailer axle must have at least 4/32 inch tread remaining. Who is correct?

    A. A only
    B. B only
    C. Both A and B
    D. Neither A nor B

42. Technician A says that EGR is an emissions device. Technician B says that an inoperative PCV can cause emission problems. Who is correct?

    A. A only
    B. B only
    C. Both A and B
    D. Neither A nor B

43. In what position should the height control valve lever be in when a truck with an air spring suspension system is at its normal height and the valve is in its neutral position?

    A. About 45 degrees upward from level
    B. Parallel to the ground
    C. About 45 degrees downward from level
    D. Approximately vertical

44. What type of lubricant should be used to grease the fifth wheel?

    A. 80/90 gear lube
    B. Water-resistant lithium-base grease
    C. Any type of bearing grease
    D. 85/140 axle lube

45. Technician A says that saturated brake linings should be replaced. Technician B says that the friction rating of linings is coded by letter codes. Who is correct?

    A. A only
    B. B only
    C. Both A and B
    D. Neither A nor B

46. What system could use a sliding gate valve?

    A. Cooling system
    B. Exhaust brake system
    C. A/C system
    D. Electrical system

47. Technician A says that spoke wheels generally experience greater alignment and balance problems than disc wheels. Technician B says that cracked wheels can be caused by vehicle overloading. Who is correct?

    A. A only

    B. B only

    C. Both A and B

    D. Neither A nor B

48. Technician A says that an improper ground could be the source of dim lights. Technician B says that continuity in an electrical circuit can be tested with a digital voltmeter/ohmmeter. Who is correct?

    A. A only

    B. B only

    C. Both A and B

    D. Neither A nor B

2013 © Delmar, Cengage Learning

49. What is shown in the illustration above?

    A. Cooling system tester

    B. Belt tension gauge

    C. Vacuum gauge

    D. Dial indicator

50. Technician A says that each time the air compressor unloader cycles, it purges water and oil residue collected in the air dryer. Technician B says that air dryers discharge air and moisture into a holding tank. Who is correct?

    A. A only

    B. B only

    C. Both A and B

    D. Neither A nor B

PREPARATION EXAM 6

1. Technician A says that a battery with a stabilized open circuit voltage of 12.2 V is about 25 percent charged. Technician B says that a battery with a stabilized open circuit voltage of 12.4 V is 75 percent charged. Who is correct?

   A. A only
   B. B only
   C. Both A and B
   D. Neither A nor B

2. When a self-adjusting clutch is found to be out of adjustment, check for all of the following EXCEPT:

   A. Correct actuator arm setting.
   B. Bent adjuster arm.
   C. Frozen adjusting ring.
   D. Clutch shaft runout.

3. Technician A says moisture and corrosion on battery tops cause faster battery self-discharge. Technician B says excessive corrosion on the battery box and cover indicates undercharging. Who is correct?

   A. A only
   B. B only
   C. Both A and B
   D. Neither A nor B

4. When testing the operation of the trailer control valve, a technician should do which of the following?

   A. Move the handle to the fully applied position and record the air pressure.
   B. Drain the air system to 0 psi, and then allow pressure to build up again.
   C. Record the gauge reading of the air pressure first, and then drain the system.
   D. Listen for air leakage around the handle as the handle is moved to the applied position.

5. In winterizing a vehicle, Technician A will use a Teflon lubricant on door latches and dry graphite on the door locks. Technician B mounts alcohol evaporator downstream from the air dryer. Who is correct?

   A. A only
   B. B only
   C. Both A and B
   D. Neither A nor B

6. Oil that has a milky, clouded appearance is probably contaminated with which of the following?

   A. Fuel
   B. Engine coolant
   C. Dust
   D. Air

7. Technician A says that wheel lug nuts should always be torqued. Technician B says that air pressure typically doesn't need to be rechecked once the tires are inflated to the proper air pressure. Who is correct?

  A. A only

  B. B only

  C. Both A and B

  D. Neither A nor B

8. All of the following could cause sluggish service brake release EXCEPT:

  A. Larger than specified brake chamber diameter.

  B. Weak chamber return spring.

  C. Broken or weak brake return springs.

  D. Obstructed brake chamber air passage.

9. Technician A says that pour point refers to the temperature at which the normal paraffins in the fuel become less soluble and begin to precipitate as wax crystals. Technician B says that the cloud point is generally 5°F to 25°F (3°C to 15°C) below the pour point. Who is correct?

  A. A only

  B. B only

  C. Both A and B

  D. Neither A nor B

10. A transmission temperature gauge does not operate. Technician A says that the sensor should be replaced because he sees the same problem recur frequently. Technician B reads a value of 1 volt at the electrical connector for the sensor and replaces the sensor because the wiring must be intact. Who is correct?

  A. A only

  B. B only

  C. Both A and B

  D. Neither A nor B

11. Filters installed in the HVAC air delivery system are:

  A. Always made of fiberglass mesh to resist corrosion.

  B. Designed to remove moisture from cab and sleeper air.

  C. Not individually replaceable.

  D. Designed to remove dust and dirt from cab and sleeper air.

12. Technician A says air seat spring suspensions eliminate the need for cab shock absorbers. Technician B says you can adjust the leveling valve to maintain the proper cab height. Who is correct?

  A. A only

  B. B only

  C. Both A and B

  D. Neither A nor B

13. When testing the maximum output of an alternator, what should the engine be run at?

    A. 1,000 rpm
    B. 1,500–2,000 rpm
    C. Idle speed
    D. 2,000 alternator rpm

14. The seat belts in the cab and sleeper are being inspected. Technician A says with the seat belt on and the truck moving, activate the parking brake to stop the truck. Technician B says it is not necessary to check the seat belts or restraints in the sleeper, only the cab. Who is correct?

    A. A only
    B. B only
    C. Both A and B
    D. Neither A nor B

15. What is the minimum performance (psi/sec) a compressor should have?

    A. 85–100 psi in 25 seconds or less
    B. 60–80 psi in 20 seconds or less
    C. 90–110 psi in 30 seconds or less
    D. 50–110 psi in 25 seconds or less

16. Technician A says that an oil sample should be taken once the oil is drained from the engine. Technician B says that an oil sample can test for contamination, metal, and dirt content. Who is correct?

    A. A only
    B. B only
    C. Both A and B
    D. Neither A nor B

17. A check of the general condition of the alternator is being performed. Technician A says to check for out-of-round mounting holes. Technician B checks the belts before making performance tests. Who is correct?

    A. A only
    B. B only
    C. Both A and B
    D. Neither A nor B

18.　In the power window circuit in the illustration above, there is an open circuit in wire 164 between the master switch and the LH power window motor. The result of this defect is:

　　A.　The LH window will go down, but it will not go back up.

　　B.　The RH window will go down, but it will not go back up.

　　C.　The LH window will stall part way down.

　　D.　The LH window is inoperative.

19. During brake inspection, a technician tests the air supply system and finds that the buildup is slow. Technician A says the compressor outlet hose may be restricted. Technician B says there may be a leak in one of the brake chambers. Who is correct?

    A. A only
    B. B only
    C. Both A and B
    D. Neither A nor B

20. A tractor with air-spring suspension sits low when coupled to a trailer and sits high without the trailer. Technician A says a leaking air spring could be the problem. Technician B says an improperly adjusted height control valve could be the problem. Who is correct?

    A. A only
    B. B only
    C. Both A and B
    D. Neither A nor B

21. The normal system pressure for a cooling system is in the range of:

    A. 8 psi.
    B. 12 psi.
    C. 15 psi.
    D. 5 psi.

22. Technician A says that a minor exhaust leak under the cab is cause for deadline of the vehicle under Commercial Vehicle Safety Alliance (CVSA) criteria. Technician B says that if the exhaust contact with air lines or any wiring is permissible as long as damage is not present. Who is correct?

    A. A only
    B. B only
    C. Both A and B
    D. Neither A nor B

23. A truck's clutch is being measured for its free travel. Technician A says the measurement should be between 1 and 2 inches. Technician B says that on a hydraulic clutch, the measurement should be 3 to 4 inches. Who is correct?

    A. A only
    B. B only
    C. Both A and B
    D. Neither A nor B

24. A driveshaft is being inspected by a technician during a PMI. Technician A says the transmission output shaft and rear-axle input shaft do not need to be checked. Technician B always checks the U-joint trunnions for cracks, the tubes for dents, and the center bearing for damage. Who is correct?

    A. A only
    B. B only
    C. Both A and B
    D. Neither A nor B

25. Technician A says that typical air filter restriction specifications are 10 inches of water vacuum for boosted engines. Technician B says that typical air filter restrictions are 30 inches of water vacuum for naturally aspirated engines. Who is correct?

    A. A only
    B. B only
    C. Both A and B
    D. Neither A nor B

26. Which leak detection method is the least accurate?

    A. Electronic leak detectors
    B. Ultraviolet (UV) leak detectors
    C. Fluid leak detectors
    D. Soapsuds solution

27. Technician A says that a voltage drop test can be used to check the condition of battery cables. Technician B says that resistance generates heat, which further increases resistance. Who is correct?

    A. A only
    B. B only
    C. Both A and B
    D. Neither A nor B

28. An exhaust system on a cab-over is being inspected. Technician A says to inspect the system while it is hot, so proper pipe to body/frame clearances may be measured. Technician B uses a prybar to check the looseness of the piping and the strength of the brackets when performing an exhaust system PMI. Who is correct?

    A. A only
    B. B only
    C. Both A and B
    D. Neither A nor B

29. A truck with an air brake system has reduced braking force on one rear wheel. Which of the following is the LEAST LIKELY cause?

    A. Restricted airflow or low air pressure to the service brake chamber
    B. Improper adjustment of slack adjuster and brake chamber pushrod
    C. Grease on shoe linings/faces
    D. A ruptured brake chamber diaphragm

30. Technician A says that alternator output should be within 20 percent of the specified rating. Technician B says that voltage output of the alternator is measured with the engine running. Who is correct?

    A. A only
    B. B only
    C. Both A and B
    D. Neither A nor B

31. Technician A says that the best method for checking a steering system is with the wheels on the shop floor and with the engine running. Technician B says that manual steering wheel play should not exceed 4 inches. Who is correct?

  A. A only

  B. B only

  C. Both A and B

  D. Neither A nor B

32. All of the statements about air brake supply systems are true EXCEPT:

  A. The supply reservoir tends to collect more moisture than the front or rear axle reservoirs.

  B. Water must be drained daily from the reservoirs if an air dryer is used in the air brake system.

  C. A small amount of oil in the supply reservoir is a normal condition.

  D. Some air dryers and automatic moisture ejectors contain electric heaters to prevent moisture from freezing in these components.

33. Technician A says that a load test should be performed after the state of charge test. Technician B says that the state of charge of a fully charged battery will be 12.4 volts. Who is correct?

  A. A only

  B. B only

  C. Both A and B

  D. Neither A nor B

2013 © Delmar, Cengage Learning

34. What is the component marked "B" in the illustration above?

  A. S-cam

  B. Power spring

  C. Check valve

  D. Pressure plate

35. Technician A says that when disconnecting a battery, the ground cable should always be removed first. Technician B says that when reconnecting a battery, the positive cable should always be connected first. Who is correct?

    A. A only

    B. B only

    C. Both A and B

    D. Neither A nor B

36. Technician A says that an inspection of the road speed governor is required during the pre-trip inspection. Technician B says that the road speed governor is not programmable. Who is correct?

    A. A only

    B. B only

    C. Both A and B

    D. Neither A nor B

2013 © Delmar, Cengage Learning

37. The slack adjuster in the illustration above should be adjusted so the brake chamber pushrod is:

    A. At an 85-degree angle in relation to the slack adjuster arm when fully applied.

    B. Parallel to the slack adjuster arm when the brakes are fully applied.

    C. At a 95-degree angle in relation to the slack adjuster arm when fully applied.

    D. At a 90-degree angle in relation to the slack adjuster arm when fully applied.

38. While inspecting a manual transmission for leaks, the technician notices that some gaskets appear to have blown out of their mating surfaces. What should the technician check first?

    A. The transmission breather

    B. The shift housing cover

    C. The shift tower

    D. The rear seal

39. A tire has a cut in the sidewall that show exposed tire cords. Technician A says that the cut should be patched. Technician B says that the tire must be replaced. Who is correct?

    A. A only
    B. B only
    C. Both A and B
    D. Neither A nor B

40. Technician A says that a loose steering gear mount could be binding the steering column and cause the power steering to become increasingly harder to turn. Technician B says that power steering wheel play should not exceed 2 ½ inches. Who is correct?

    A. A only
    B. B only
    C. Both A and B
    D. Neither A nor B

41. All of the following practices should be avoided when working with truck frames EXCEPT:

    A. Drilling holes in the frame web.
    B. Drilling holes in the frame flanges.
    C. Cutting holes in the frame with a torch.
    D. Notching the frame rails.

42. Technician A says that a milky gray sludge indicates water in the oil. Technician B says that used oil and filters are considered hazardous waste. Who is correct?

    A. A only
    B. B only
    C. Both A and B
    D. Neither A nor B

43. When testing a suspension system on a heavy-duty truck, a technician should do which of the following?

    A. Bounce the vehicle up and down to test the shocks.
    B. Visually inspect the suspension components for signs of fatigue, deterioration, and/or failure.
    C. Road test the vehicle to determine possible defects and/or misalignment(s).
    D. Check the PMI schedule for scheduled repairs.

44. Technician A says that the purpose of the radiator cap is to allow lower operating temperatures. Technician B says that an underperforming radiator cap may cause engine overheating and coolant loss. Who is correct?

    A. A only
    B. B only
    C. Both A and B
    D. Neither A nor B

45. A slack adjuster is being adjusted. Technician A says excessive wear limits between the clevis and the collar will cause the stroke to be too long. Technician B says improper brake adjustment may cause premature wheel lockup during a brake application. Who is correct?

    A. A only

    B. B only

    C. Both A and B

    D. Neither A nor B

46. A gauge cluster is being inspected for proper function and illumination. Technician A says that a malfunctioning oil pressure gauge is not a valid reason for tagging a vehicle out of service. Technician B says that a proper function check of the instrumentation panel is required under a PMI. Who is correct?

    A. A only

    B. B only

    C. Both A and B

    D. Neither A nor B

47. All of the following spring types can be found on a heavy-duty truck suspension EXCEPT:

    A. Constant rate.

    B. Progressive rate.

    C. Auxiliary.

    D. Coil.

Brakes applied

Brakes released

Brake shoes

Brake shoe rollers   S-cam

2013 © Delmar, Cengage Learning

48. The brake linings are worn and there is excessive clearance between the linings and the brake drum on a cam-type foundation brake (see illustration above). Technician A says this condition may cause the cam to roll over during a brake application. Technician B says this condition may cause loss of steering control. Who is correct?

A. A only

B. B only

C. Both A and B

D. Neither A nor B

49. Technician A says that black smoke is normal and not a cause for concern. Technician B says black smoke is a sign that the engine is getting the correct amount of fuel and air for optimum performance. Who is correct?

A. A only

B. B only

C. Both A and B

D. Neither A nor B

50. Technician A says vibration damper absorbs camshaft harmonic distortions. Technician B says the vibrations damper balances harmonic vibrations caused by torsional loads of the rotating crankshaft. Who is correct?

A. A only

B. B only

C. Both A and B

D. Neither A nor B

6 Answer Keys and Explanations

INTRODUCTION

Included in this section are the answer keys for each preparation exam, followed by individual, detailed answer explanations and a reference identifying the designated task area being assessed by each specific question. This additional reference information may prove useful if you need to refer back to the task list located in Section 4 of this book for additional support.

PREPARATION EXAM 1—ANSWER KEY

1.	B	21.	D	41.	B
2.	D	22.	D	42.	D
3.	D	23.	A	43.	D
4.	C	24.	C	44.	A
5.	B	25.	A	45.	C
6.	C	26.	C	46.	A
7.	A	27.	D	47.	D
8.	B	28.	B	48.	B
9.	C	29.	D	49.	B
10.	A	30.	C	50.	A
11.	A	31.	B		
12.	C	32.	C		
13.	D	33.	A		
14.	C	34.	B		
15.	B	35.	B		
16.	C	36.	C		
17.	A	37.	C		
18.	D	38.	A		
19.	B	39.	C		
20.	D	40.	A		

PREPARATION EXAM 1—EXPLANATIONS

1. The fifth wheel to upper coupler connection should have no more than how many inches of longitudinal play?

    A. 1/16

    B. 1/8

    C. 1/4

    D. 1/2

    TASK D.**4.8**

    Answer A is incorrect. Up to 1/8 inch is acceptable.

    Answer B is correct. 1/8 inch play is acceptable freeplay.

    Answer C is incorrect. 1/4 inch is too much play.

    Answer D is incorrect. 1/2 inch is too much play.

2. When a starting motor is removed, there are shims positioned between the starting motor mounting flange and the flywheel housing. Technician A says these shims provide improved electrical contact between the starting motor and the flywheel housing. Technician B says these shims provide easier installation of the starting motor mounting bolts. Who is correct?

    TASK C.**5**

    A. A only

    B. B only

    C. Both A and B

    D. Neither A nor B

    Answer A is incorrect. The shims between the starting motor flange and the flywheel housing set the correct depth of the starter drive pinion to the flywheel ring gear.

    Answer B is incorrect. The shims between the starting motor flange and the flywheel housing provide proper starting motor alignment.

    Answer C is incorrect. Neither Technician is correct.

    Answer D is correct. Neither Technician is correct. The shims between the starting motor flange and the flywheel housing provide proper starting motor alignment.

2013 © Delmar, Cengage Learning

TASKS D.3.7,
D.3.9

3. What is the operation being performed in the illustration above?

   A.   Checking wheel-hub runout

   B.   Aligning wheel hub

   C.   Installing wheel bearings

   D.   Installing a wheel seal

Answer A is incorrect. Wheel hub runout is not being checked.

Answer B is incorrect. A wheel is not aligned to an axle hub.

Answer C is incorrect. Wheel bearings are usually installed by hand.

Answer D is correct. The illustration shows the installation of front wheel bearing seals.

TASKS B.4,
B.5

4. Technician A says that the blower motor should operate in all speed positions. Technician B says idle shutdown duration is programmable on electronic engines. Who is correct?

   A.   A only

   B.   B only

   C.   Both A and B

   D.   Neither A nor B

Answer A is incorrect. Technician B is also correct.

Answer B is incorrect. Technician A is also correct.

Answer C is correct. Both Technicians are correct. The blower motor should operate in all speed positions. Idle shutdown duration is a programmable feature on most electronically controlled engines. Engine shutdown can be programmed to occur after a desired time interval. When the desired time has elapsed without input from the throttle pedal or clutch pedal, the engine will shut down automatically. To change this programmed time, an EST is needed.

Answer D is incorrect. Both Technicians are correct.

5. Technician A says that you can mix radial and bias tires on a steering axle as long as the vehicle doesn't exceed 55 mph. Technician B says that tires of different tread design mounted on the same axle are a deadline condition in many jurisdictions. Who is correct?

TASK D.3.18

   A. A only

   B. B only

   C. Both A and B

   D. Neither A nor B

Answer A is incorrect. You cannot mix radial tires and bias tires on a steering axle.

Answer B is correct. Tires of different tread design mounted on the same axle are a deadline condition in many jurisdictions.

Answer C is incorrect. Only Technician B is correct.

Answer D is incorrect. Technician B is correct.

6. If the fuel is contaminated with water, the water will:

TASK A.10

   A. Remain mixed with the fuel.

   B. Float on top of the fuel.

   C. Settle to the bottom of the fuel tank.

   D. Flow through the filters into the injection system.

Answer A is incorrect. Water does not remain mixed with the fuel.

Answer B is incorrect. Water does not float on top of the fuel.

Answer C is correct. Water settles to the bottom of the fuel tank.

Answer D is incorrect. Water usually does not flow through the filters.

7. Technician A says the back-up warning system provides a warning to anyone standing behind the truck. Technician B says the back-up warning system provides a visual warning. Who is correct?

TASK E.10

   A. A only

   B. B only

   C. Both A and B

   D. Neither A nor B

Answer A is correct. Only Technician A is correct. The back-up warning system provides a warning to anyone standing behind the truck.

Answer B is incorrect. The back-up warning system provides an audible warning.

Answer C is incorrect. Only Technician A is correct.

Answer D is incorrect. Technician A is correct.

8. A binding steering wheel may be caused by:

TASKS D.3.1, E.3

   A. Loose tie rod ends.

   B. Loose or worn steering shaft U-joints.

   C. Loose drag-link ends.

   D. Worn kingpin bushings.

Answer A is incorrect. Loose tie rod ends do not cause a binding steering wheel.

Answer B is correct. Loose or worn steering shaft U-joints may cause a binding steering wheel.

Answer C is incorrect. Loose drag-link ends do not cause a binding steering wheel.

Answer D is incorrect. Worn kingpin bushings do not cause a binding steering wheel.

TASK C.8

9. Technician A says that an inductive clamp ammeter could be used to check alternator output. Technician B says that alternator output decreases as temperature increases. Who is correct?

    A. A only
    B. B only
    C. Both A and B
    D. Neither A nor B

    Answer A is incorrect. Technician B is also correct.

    Answer B is incorrect. Technician A is also correct.

    Answer C is correct. Both Technicians are correct. You can use an inductive clamp ammeter to check alternator output. It is normal for alternator output to slightly decrease as ambient temperature increases.

    Answer D is incorrect. Both Technicians are correct.

TASK D.1.5

10. An audible air leak is observed that appears to be coming from the front brake service brake chamber with the brakes applied. Which of the following is the LEAST LIKELY cause?

    A. Loose mounting bolts
    B. Ruptured diaphragm
    C. Loose chamber clamp
    D. Missing clamp bolt

    Answer A is correct. Loose mounting bolts would cause the chamber to move when the brakes are applied, but would be unlikely to cause an audible leak.

    Answer B is incorrect. A ruptured diaphragm could cause an air leak.

    Answer C is incorrect. A loose chamber clamp could cause an air leak.

    Answer D is incorrect. A missing clamp bolt could cause an air leak.

TASK A.14

11. During a preventive maintenance inspection (PMI), Technician A says the air passages through the air-to-air intercooler should be checked for restrictions. Technician B says the air-to-air intercooler uses refrigerant to lower the intake air temperature. Who is correct?

    A. A only
    B. B only
    C. Both A and B
    D. Neither A nor B

    Answer A is correct. Only Technician A is correct. The air passages through the air-to-air intercooler should be checked for restrictions.

    Answer B is incorrect. The air-to-air intercooler does not use refrigerant to lower intake air temperature.

    Answer C is incorrect. Only Technician A is correct.

    Answer D is incorrect. Technician A is correct.

12. If a starter ground circuit test reveals a voltage drop of more than 0.5 volts, which of the following may be the cause of the problem?

    A. A shorted armature winding
    B. An open field coil
    C. A damaged battery ground cable
    D. A faulty ignition switch

TASK C.2

Answer A is incorrect. A shorted armature winding would cause increased current flow in the insulated side of the circuit.

Answer B is incorrect. An open field coil would stop or interrupt current flow on the insulated side.

Answer C is correct. A damaged or corroded battery ground cable could show a voltage drop of more than 0.5 volt. Many cranking motors and batteries are needlessly replaced due to defective or corroded cables and terminals. A voltage drop value exceeding 0.2 volt through any individual terminal or cable indicates a possible problem. Electrical connections should not produce voltage drops greater than 0.1 volt and total voltage drop of all the cables and terminals in a cranking circuit should not exceed 0.5 volt.

Answer D is incorrect. The ignition-switch circuit is not related to the ground-circuit resistance test.

13. The most likely cause of blue smoke in diesel exhaust is:

    A. Coolant entering the combustion chambers.
    B. A restricted intake.
    C. Low fuel pressure to the injectors.
    D. Oil entering the combustion chambers.

TASK E.7

Answer A is incorrect. Coolant entering the combustion chambers causes white or gray smoke in the exhaust.

Answer B is incorrect. A restricted intake may cause black smoke in the exhaust, but not blue smoke.

Answer C is incorrect. Low fuel pressure does not cause blue smoke in the exhaust.

Answer D is correct. Oil entering the combustion chambers causes blue smoke in the exhaust.

2013 © Delmar, Cengage Learning

14. What does the gauge in the illustration above show?

    A. Water temperature
    B. Oil temperature
    C. Oil pressure
    D. Transmission temperature

TASK B.3

Answer A is incorrect. A water temperature gauge would have different values and symbol(s).

Answer B is incorrect. An oil temperature gauge would have different values and symbol(s).

Answer C is correct. An oil pressure gauge is shown.

Answer D is incorrect. A transmission temperature gauge would have different values and symbols.

TASK D.1.26

15. When adjusting hydraulic drum brakes, Technician A says to use a pry bar to center the brake shoes. Technician B says to turn the manual adjusting wheel (star wheel) to adjust shoe clearance. Who is correct?

  A. A only
  B. B only
  C. Both A and B
  D. Neither A nor B

Answer A is incorrect. Apply and release the brake pedal several times to center the brake shoes.

Answer B is correct. Only Technician B is correct. If the shoe clearance is not correct, turn the manual adjusting wheel (star wheel) on the wheel cylinder to obtain the correct clearance.

Answer C is incorrect. Only Technician B is correct.

Answer D is incorrect. Technician B is correct.

TASK D.2.19

16. Technician A says drive axle oil samples are taken for analysis of lubricating qualities, metal, and dirt content. Technician B says that oil samples are taken to check for corrosion-causing agents. Who is correct?

  A. A only
  B. B only
  C. Both A and B
  D. Neither A nor B

Answer A is incorrect. Technician B is also correct.

Answer B is incorrect. Technician A is also correct.

Answer C is correct. Both Technicians are correct. During a PMI, transmission and drive axle oil samples are taken for analysis of lubricating qualities, metal, and dirt content. Also, corrosion-causing agents such as sulfuric and hydrochloric acids are tested.

Answer D is incorrect. Both Technicians are correct.

TASK A.17

17. A gasoline engine experiences rough idle operation. This problem is worse when the engine is at normal operating temperature. The cylinder compression is within specifications and all the ignition components are satisfactory. Which of the following could be the cause of this problem?

  A. An exhaust gas recirculation (EGR) valve stuck open
  B. A burned exhaust valve
  C. Valve adjustment that is too tight
  D. A bent intake valve

Answer A is correct. An EGR valve stuck open causes rough idle operation.

Answer B is incorrect. A burned exhaust valve causes low compression.

Answer C is incorrect. A valve adjustment that is too tight causes low cylinder compression.

Answer D is incorrect. A bent intake valve causes low cylinder compression.

TASK A.15

18. A turbocharger on a tractor has less than the specified boost pressure. All of these defects may be the cause of the problem EXCEPT:

  A. Turbocharger bearings that are starting to fail.
  B. Improper wastegate adjustment.
  C. Damaged compressor wheel.
  D. Turbocharger oil seals leaking oil into the exhaust.

Answer A is incorrect. Turbocharger bearings that are starting to fail may cause reduced boost pressure.

Answer B is incorrect. Improper wastegate adjustment may cause reduced boost pressure.

Answer C is incorrect. A damaged compressor wheel may cause reduced boost pressure.

Answer D is correct. Turbocharger oil seals leaking oil into the exhaust cause blue smoke in the exhaust, but this problem does not cause reduced boost pressure.

19. A wheel cylinder:

    A. Does not need any maintenance.
    B. Needs to be inspected at a PMI.
    C. Needs to be replaced periodically.
    D. Is usually constructed of aluminum.

TASK D.1.24

Answer A is incorrect. A hydraulic wheel cylinder may occasionally need to be rebored.

Answer B is correct. A wheel cylinder should always be inspected at a PMI.

Answer C is incorrect. A wheel cylinder does not need replacing periodically, but needs servicing on occasion.

Answer D is incorrect. Wheel cylinders are generally constructed of cast iron.

20. Technician A says that it is OK to use ether on glow-plug–equipped engines. Technician B always uses ether when starting a diesel engine. Who is correct?

    A. A only
    B. B only
    C. Both A and B
    D. Neither A nor B

TASK A.2

Answer A is incorrect. Using ether on glow-plug-equipped engines can cause serious internal engine damage.

Answer B is incorrect. Ether is generally used only in cold weather conditions and should not be needed in warmer weather.

Answer C is incorrect. Neither Technician is correct.

Answer D is correct. Neither Technician is correct. Ether should never be used on a glow-plug-equipped engine. Ether is generally used only in cold weather conditions.

21. While discussing alternator output testing, Technician A says when the alternator is full-fielded, the voltage should be kept below 17 V. Technician B says the output may be tested by lowering the system voltage to 11.5 V. Who is correct?

    A. A only
    B. B only
    C. Both A and B
    D. Neither A nor B

TASK C.8

Answer A is incorrect. When the alternator is full-fielded, the voltage should be kept below 15 V.

Answer B is incorrect. When testing output by lowering the system voltage, the voltage should be lowered to 13 V.

Answer C is incorrect. Neither Technician is correct.

Answer D is correct. Neither Technician is correct. When the alternator is full-fielded, the voltage should be kept below 15 V. When testing output by lowering the system voltage, the voltage should be lowered to 13 V.

TASK C.11

22. What would occur if the circuit, indicated at point X in the illustration above, was found to be heavily corroded?

   A. The right headlight would be inoperative.

   B. The left headlight would be inoperative.

   C. The right high beam would be dim.

   D. The left high beam would be dim.

Answer A is incorrect. Power going to the right headlight would not be interrupted.

Answer B is incorrect. Power going to the left headlight would not be interrupted.

Answer C is incorrect. Power going to the right high beam would be unaffected.

Answer D is correct. The left high-beam headlight would be dim. Corrosion causes unwanted high circuit resistance, causing a decrease in current flow. Most electrical loads have a specific resistance designed into them to operate at a predicted voltage and current level. Any unwanted resistance in the circuit will affect the load device.

23. Technician A says that an inspection of the road speed governor is required during the road/operational test. Technician B says that more emphasis should be placed on the post-trip inspection rather than the pre-trip inspection. Who is correct?

TASK E.5

   A. A only
   B. B only
   C. Both A and B
   D. Neither A nor B

   Answer A is correct. Only Technician A is correct. The operation of the road speed governor is checked during the road/operational test.

   Answer B is incorrect. A post-trip inspection is of secondary importance.

   Answer C is incorrect. Only Technician A is correct.

   Answer D is incorrect. Technician A is correct.

24. When inspecting an oil filter housing or its mounting, which of the following is LEAST LIKELY to be a procedure?

TASK A.27

   A. Visually checking for cracks
   B. Visually checking gasket surface for nicks
   C. Magna fluxing the housing and metal parts to detect small cracks
   D. Visually inspecting housing passageways for obstruction and cracks

   Answer A is incorrect. A visual inspection for cracks is a must.

   Answer B is incorrect. A visual inspection of the gasket surface will help ensure a tight leak-free seal.

   Answer C is correct. Magna fluxing of housing assembly and metal parts to detect small cracks would typically not be performed.

   Answer D is incorrect. A visual inspection of all passageways is necessary to ensure oil flows.

25. What is the most likely mileage recommended in changing synthetic lubricant for truck drive axles driven in a linehaul application?

TASK D.2.15

   A. 500,000 miles
   B. 200,000 miles
   C. 100,000 miles
   D. 50,000 miles

   Answer A is correct. 500,000 miles is the most likely change interval for class "A" highway vehicles that use synthetic lubricant.

   Answer B is incorrect. 200,000 miles is not a suitable interval for changing axle lubricant.

   Answer C is incorrect. 100,000 miles is not a suitable interval for changing axle lubricant.

   Answer D is incorrect. 50,000 miles is not enough mileage for an axle lubricant change.

TASKS B.12,
B.13, B.14

26. Technician A says that a vehicle must have proof of inspection according to the Federal Highway Administration. Technician B says that any wiper unit that has inoperable parts on the driver's side shall be cause for deadline according to CVSA safety standards. Who is correct?

A. A only

B. B only

C. Both A and B

D. Neither A nor B

Answer A is incorrect. Technician B is also correct.

Answer B is incorrect. Technician A is also correct.

Answer C is correct. Both Technicians are correct. The Federal Highway Administration has set up a minimum inspection standards program, under which each vehicle must carry proof that an inspection was completed. Proof can either be a copy of the inspection form kept on the vehicle or a decal. If using a decal, a copy of the inspection form must be kept on file and must indicate where an inspector can get a copy of it. The only requirement of a decal itself is that it be legible. Each vehicle must be inspected separately. CVSA safety standards out-of-service criteria state that any wiper unit that has inoperable parts or missing wipers that render the system ineffective on the driver's side shall be cause for deadline.

Answer D is incorrect. Both Technicians are correct.

TASKS D.1.3,
D.1.9, D.1.10

27. What is not checked when inspecting an air brake system?

A. Tractor protection valve

B. Spring brake inversion system

C. Desiccant pack in air dryer

D. Operation of the clutch brake

Answer A is incorrect. A tractor protection valve check is always part of an air brake system check.

Answer B is incorrect. Proper operation of the spring brake inversion system is vital.

Answer C is incorrect. Non-regenerative desiccant packs must be replaced regularly.

Answer D is correct. Checking the operation of the clutch brake is not part of the brakes. It is in the clutch unit.

TASK C.10

28. A heavy-duty diesel-powered truck is having its electrical system inspected. The interior lights glow dimly even at a high engine speed. Technician A says that a defective ballast resistor should be suspected. Technician B says to check the cab ground integrity. Who is correct?

A. A only

B. B only

C. Both A and B

D. Neither A nor B

Answer A is incorrect. Ballast resistors are used for ignition systems.

Answer B is correct. Only Technician B is correct. You should check the cab ground integrity. Interior cab lights and circuits are affected by resistance problems such as loose terminals, corroded connectors, and high-resistance ground connections. Proper grounding of the dash circuits and interior lighting should always be checked whenever a problem is suspected. When more than one light has similar symptoms, like not operating or being dim, the cab may not be adequately grounded. This is especially true when lights are attached to plastic panels, and the ground must be completed elsewhere.

Answer C is incorrect. Only Technician B is correct.

Answer D is incorrect. Technician B is correct.

To HR-1
relay valve

2013 © Delmar, Cengage Learning

29. What is indicated in the illustration above?

   A. Hydraulic liftgate cylinder

   B. By-pass solenoid for reefer unit

   C. Remote automatic kingpin locking relay

   D. Hydraulic spring parking brake actuator

TASK D.1.22

Answer A is incorrect. A hydraulic liftgate cylinder would be much larger.

Answer B is incorrect. The indicated object is not a solenoid.

Answer C is incorrect. The kingpin does not have an automatic locking relay.

Answer D is correct. A hydraulic spring parking brake actuator is shown in the illustration.

30. While discussing water pumps, Technician A says a defective water pump bearing may cause a growling noise with the engine idling. Technician B says the water pump bearing may be ruined by coolant leaking past the water pump seal. Who is correct?

   A. A only

   B. B only

   C. Both A and B

   D. Neither A nor B

TASK A.26

Answer A is incorrect. Technician B is also correct.

Answer B is incorrect. Technician A is also correct.

Answer C is correct. Both Technicians are correct. A defective water pump bearing may cause a growling noise with the engine idling. The water pump bearing may be ruined by coolant leaking past the water pump seal.

Answer D is incorrect. Both Technicians are correct.

31. Cloudiness in hydraulic brake fluid is a sign of which of the following?

   A. Air in the fluid

   B. Dirt contamination

   C. Overheated/burnt fluid

   D. Cold fluid

TASK D.1.20

Answer A is incorrect. Aerated fluid will appear with small bubbles trapped within it.

Answer B is correct. Dirt or rust contaminated brake fluid will appear cloudy or opaque.

Answer C is incorrect. Burned or otherwise overheated brake fluid will have a distinct and pungent odor of being unlike any other brake fluid, such that a distinction can be made.

Answer D is incorrect. Brake fluid will not freeze above the working limits of a vehicle.

TASKS C.3,
C.4

32. A battery is being tested on a truck during PMI inspection. Technician A says that the surface charge should be removed from a battery before it can be load tested. Technician B says when load testing, you should apply a load equal to half of the battery's cold cranking amperes (CCA) rating for 15 seconds. Who is correct?

A.　A only

B.　B only

C.　Both A and B

D.　Neither A nor B

Answer A is incorrect. Technician B is also correct.

Answer B is incorrect. Technician A is also correct.

Answer C is correct. Both Technicians are correct. A battery that has recently been charged by running the engine, or using a battery charger, must have its surface charge removed before performing a battery load test. The surface charge may be removed by applying a load of 300 amps for 15 seconds with a carbon pile load tester, or by cranking the engine for 15 seconds. After removing the surface charge, stabilize the voltage by allowing the battery to rest for 15 minutes. If multiple batteries connected, then first disconnect the batteries and perform individual testing on each battery. If the voltage is below 12.4 volts DC at 70°F, the battery should be recharged to above 12.6 volts. Apply a load through the tester at one-half the CCA for 15 seconds. If the voltage drops below 9.6 volts at 70°F, the battery is unserviceable.

Answer D is incorrect. Both Technicians are correct.

TASK D.1.23

33. All of the following problems may cause a low, spongy brake pedal EXCEPT:

A.　Excessive brake pedal freeplay.

B.　Air in the hydraulic brake system.

C.　Brake fluid contaminated with moisture.

D.　Low brake fluid level.

Answer A is correct. Excessive brake pedal freeplay causes a low, firm pedal.

Answer B is incorrect. Air in the hydraulic system causes a low, spongy brake pedal.

Answer C is incorrect. Brake fluid contaminated with moisture causes a low, spongy brake pedal.

Answer D is incorrect. Low brake fluid level causes a low, spongy brake pedal.

TASK D.3.17

34. What do cracks between wheel lug holes indicate?

A.　Over inflation of tires

B.　Overloading

C.　Using larger than recommended tire sizes

D.　Incorrect wheel size for application

Answer A is incorrect. Over inflation causes cracks in the rim base.

Answer B is correct. Overloading causes lug hole cracking.

Answer C is incorrect. Using larger than recommended tires causes cracks in the rim base.

Answer D is incorrect. Incorrect wheel size generally does not cause cracks.

35. Technician A says that when filling the cooling system, the air bleeder valves should be closed. Technician B says that if the coolant solution contains over 67 percent antifreeze, heat transfer from the cylinders to the coolant is adversely affected. Who is correct?

    A. A only
    B. B only
    C. Both A and B
    D. Neither A nor B

TASK A.25

Answer A is incorrect. When filling the cooling system, the air bleeder valves should be open.

Answer B is correct. Only Technician B is correct. If the coolant solution contains over 67 percent antifreeze, heat transfer from the cylinders to the coolant is adversely affected.

Answer C is incorrect. Only Technician B is correct.

Answer D is incorrect. Technician B is correct.

36. When adjusting the height control valve in an air suspension system:

    A. The height control valve arm may be moved upward to lower the suspension height.
    B. Steel locating pins are installed in the height control valve arms.
    C. The air brake system pressure must be 70 psi (483 kPa) or more.
    D. The truck must have an average load on the chassis.

TASK D.4.3

Answer A is incorrect. The height control valve arm may be moved upward to raise the suspension height.

Answer B is incorrect. Wood locating pins are installed in the height control valve arm.

Answer C is correct. The air brake system pressure must be 70 psi (483 kPa) or more.

Answer D is incorrect. The truck must be unloaded or empty.

37. Technician A says a loose grab handle may cause personal injury. Technician B says the steps must be level and free of debris. Who is correct?

    A. A only
    B. B only
    C. Both A and B
    D. Neither A nor B

TASK B.18

Answer A is incorrect. Technician B is also correct.

Answer B is incorrect. Technician A is also correct.

Answer C is correct. Both Technicians are correct. A loose grab handle may cause personal injury. The steps must be level and free of debris.

Answer D is incorrect. Both Technicians are correct.

Adjuster hole

Adjuster lever push back 1/16" max (1.5 mm)

Move handle upward to retract brake shoes, downward to expand

2013 © Delmar, Cengage Learning

TASK D.1.26

38. On the hydraulic drum brake in the illustration above, the technician is:
    A. Performing a hydraulic drum brake adjustment.
    B. Changing a return spring on a hydraulic brake.
    C. Measuring brake shoe to drum clearance.
    D. Setting the park brake gap.

    Answer A is correct. The technician is performing a hydraulic drum brake adjustment.

    Answer B is incorrect. The technician is performing a hydraulic drum brake adjustment.

    Answer C is incorrect. The technician is performing a hydraulic drum brake adjustment.

    Answer D is incorrect. The technician is performing a hydraulic drum brake adjustment.

TASK C.7

39. While discussing alternator noise complaints, Technician A says a loose alternator belt may cause a squealing noise. Technician B says a shorted alternator diode may cause a whining noise. Who is correct?
    A. A only
    B. B only
    C. Both A and B
    D. Neither A nor B

    Answer A is incorrect. Technician B is also correct. A loose alternator belt may cause a squealing noise.

    Answer B is incorrect. Technician A is also correct. A shorted alternator diode may cause a whining noise.

    Answer C is correct. Both Technicians are correct.

    Answer D is incorrect. Both Technicians are correct.

TASKS D.1.15, D.1.16

40. The S-cam on one foundation brake fitted with Q brakes rolls over. Which of the following is the most likely cause?
    A. Excessively worn drums and shoes
    B. Application air pressure too high
    C. Improperly installed slack adjuster
    D. Brake chamber push rod too long

    Answer A is correct. When brake drums and shoes are excessively worn, trying to adjust them to specs will put the cam in a position to roll over when the brakes are applied.

    Answer B is incorrect. Application of air pressure affects all vehicle brakes and will not cause the S-cam to rollover.

    Answer C is incorrect. Improperly installed slack adjusters will not cause S-cam rollover.

    Answer D is incorrect. Brake chamber push rod length will not cause S-cam rollover.

41. Technician A says air seat spring suspensions eliminate the need for cab shock absorbers. Technician B says you can adjust the leveling valve to maintain the proper seat height. Who is correct?

TASKS B.16, B.26

   A. A only
   B. B only
   C. Both A and B
   D. Neither A nor B

   Answer A is incorrect. Air seat spring suspensions do not eliminate the need for cab shock absorbers.

   Answer B is correct. Only Technician B is correct. You can adjust the leveling valve to maintain the proper seat height.

   Answer C is incorrect. Only Technician B is correct.

   Answer D is incorrect. Technician B is correct.

42. The following are part of an automatic chassis lube system EXCEPT:

TASKS D.4.5, D.4.11

   A. Grease lines.
   B. Reservoir.
   C. Pump.
   D. Manifold or distribution block.

   Answer A is incorrect. Grease lines serve as avenues for the grease to travel from one place to another.

   Answer B is incorrect. Reservoirs are the components that store the medium until needed.

   Answer C is incorrect. The pump is the delivery mechanism for the system.

   Answer D is correct. The manifold or distribution block is not part of the automatic chassis lube system.

43. While discussing safety starting switches, Technician A says the safety starting switch is connected parallel to the solenoid windings. Technician B says a defective safety starting switch may cause higher than specified starter current draw. Who is correct?

TASK C.6

   A. A only
   B. B only
   C. Both A and B
   D. Neither A nor B

   Answer A is incorrect. The safety starting switch is connected in series with the solenoid windings.

   Answer B is incorrect. A defective safety starting switch does not cause higher than specified starter current draw.

   Answer C is incorrect. Neither Technician is correct.

   Answer D is correct. Neither Technician is correct. The safety starting switch is connected in series with the solenoid windings. A defective safety starting switch does not cause higher than specified starter current draw.

TASKS D.2.13,
D.4.5, D.4.11

44. What activates the pump in an air-driven automatic chassis lube system (ACLS)?

A.   A solenoid valve

B.   A relay valve

C.   An electronic control module (ECM)

D.   A toggle switch inside the cab

Answer A is correct. A solenoid valve activates the pump.

Answer B is incorrect. There are no relays in an air-driven automatic chassis lubrication system.

Answer C is incorrect. There is no ECM to control an ACLS.

Answer D is incorrect. A toggle switch would make the system manual.

TASK A.5

45. When troubleshooting a diesel engine which uses excessive amounts of oil, which of the following should be checked first?

A.   Worn oil rings

B.   Oil leakage through the engine air compressor

C.   External oil leaks

D.   Oil leakage through the turbocharger

Answer A is incorrect. Although worn oil rings will cause excessive use of oil, it wouldn't be checked first.

Answer B is incorrect. Although oil leakage through the air engine air compressor will cause excessive use of oil, it wouldn't be checked first.

Answer C is correct. External oil leaks will cause excessive use of oil and should be checked first.

Answer D is incorrect. Although worn oil leakage through the turbocharger will cause excessive use of oil, it wouldn't be checked first.

TASKS C.7,
B.2

46. The charging indicator light on a truck is illuminated when the ignition switch is turned off. Technician A says some of the alternator diodes may be defective. Technician B says the wire connected to the alternator L terminal may be grounded. Who is correct?

A.   A only

B.   B only

C.   Both A and B

D.   Neither A nor B

Answer A is correct. Only Technician A is correct. Some of the alternator diodes may be defective.

Answer B is incorrect. If the wire connected to the alternator L terminal is grounded the charge indicator light is not on with the ignition switch off.

Answer C is incorrect. Only Technician A is correct.

Answer D is incorrect. Technician A is correct.

TASK D.1.16

47. The machining limit of a brake drum is 14.060 inches, and the discard limit is:

A.   14.075 inches.

B.   14.080 inches.

C.   15.005 inches.

D.   14.090 inches.

Answer A is incorrect. The discard limit is 14.090 inches.

Answer B is incorrect. The discard limit is 14.090 inches.

Answer C is incorrect. The discard limit is 14.090 inches.

Answer D is correct. The discard limit is 14.090 inches.

48. Technician A says a transmission oil sample can be taken after the oil is drained into a drain pan. Technician B says to take a transmission oil sample mid-way through the drain-off. Who is correct?

TASK D.2.18

   A. A only
   B. B only
   C. Both A and B
   D. Neither A nor B

   Answer A is incorrect. The transmission oil sample must be taken mid-way through the drain-off.

   Answer B is correct. Only Technician B is correct. You always take a transmission oil sample mid-way through the drain-off.

   Answer C is incorrect. Only Technician B is correct.

   Answer D is incorrect. Technician B is correct.

49. Technician A says to check the oil level with the engine running. Technician B says an oil level too low can cause engine failure. Who is correct?

TASK A.7

   A. A only
   B. B only
   C. Both A and B
   D. Neither A nor B

   Answer A is incorrect. The recommended procedure for checking the oil level is to ensure that the vehicle is on level ground and has not been operated for at least five minutes.

   Answer B is correct. Only Technician B is correct. An oil level that is too low may allow air to be drawn into the pump and a loss of oil circulation, resulting in engine failure.

   Answer C is incorrect. Only Technician B is correct.

   Answer D is incorrect. Technician B is correct.

50. When servicing the power take off (PTO) unit, which of the following is LEAST LIKELY to result in replacement of the PTO?

TASK D.2.16

   A. Missing gear teeth
   B. Leaks
   C. Cracking
   D. Galling

   Answer A is incorrect. Missing gear teeth may require replacement of the PTO unit.

   Answer B is correct. Leaks are generally easy to repair and usually do not require replacement of the PTO unit.

   Answer C is incorrect. Cracks in the PTO unit may require replacement.

   Answer D is incorrect. Pitting of the PTO unit may require replacement.

PREPARATION EXAM 2—ANSWER KEY

1.	A	18.	D	35.	B
2.	D	19.	A	36.	D
3.	D	20.	B	37.	A
4.	C	21.	D	38.	C
5.	A	22.	C	39.	C
6.	D	23.	B	40.	C
7.	B	24.	A	41.	A
8.	C	25.	D	42.	C
9.	B	26.	B	43.	D
10.	C	27.	A	44.	A
11.	A	28.	D	45.	C
12.	D	29.	A	46.	A
13.	C	30.	A	47.	A
14.	C	31.	D	48.	D
15.	C	32.	B	49.	B
16.	B	33.	C	50.	A
17.	B	34.	A		

PREPARATION EXAM 2—EXPLANATIONS

TASK C.8

1. All of the following may result in an undercharged battery EXCEPT:

   A. A defective voltage regulator causing a high-voltage regulator setting.

   B. Excessive resistance between the alternator battery terminal and the positive battery terminal.

   C. Loose or worn alternator drive belt.

   D. A 60-ampere fuse link installed in place of a 100-ampere fuse link in the alternator battery wire.

   Answer A is correct. A defective voltage regulator causing a high-voltage regulator setting results in an overcharged battery.

   Answer B is incorrect. Excessive resistance between the alternator battery terminal and the positive battery terminal may cause an undercharged battery.

   Answer C is incorrect. A loose or worn alternator drive belt can cause slippage and the alternator not turning at the correct speed resulting in an undercharged battery.

   Answer D is incorrect. A 60-ampere fuse link installed in place of a 100-ampere fuse link in the alternator battery wire could result in fuse failure resulting in an undercharged battery.

2. A technician is testing the low air pressure warning system on a truck by performing a pressure drop test on a straight single truck's air brake system. Technician A says that the pressure drop should be no more than 6 psi in two minutes. Technician B says you hold down the brake pedal for the first minute then release it. Who is correct?

    TASK D.1.7

    A. A only
    B. B only
    C. Both A and B
    D. Neither A nor B

    Answer A is incorrect. The pressure drop on a straight truck is 4 psi in two minutes.

    Answer B is incorrect. The brake pedal must be held down for the two-minute test.

    Answer C is incorrect. Neither Technician is correct.

    Answer D is correct. Neither Technician is correct. The pressure drop on a straight truck is 4 psi in two minutes. The brake pedal must be held down for the two-minute test.

3. What is the most likely cited maintenance item on transport trailers?

    TASK E.8

    A. Tires
    B. Lights
    C. Suspension
    D. Brakes

    Answer A is incorrect. Although tires are a high-wear item, they account for 17.6 percent of complaints.

    Answer B is incorrect. Lights account for 14.5 percent of all maintenance concerns.

    Answer C is incorrect. Suspension concerns account for only 6.2 percent of maintenance concerns.

    Answer D is correct. Brakes are the highest cited area on a PMI. Brakes account for 25.9 percent of maintenance concerns and top the list of maintenance concerns that fleets cite when asked to name the top five trailer maintenance problems.

4. Technician A says to connect the battery positive cable first when connecting a battery. Technician B says to disconnect the ground cable first when disconnecting a battery. Who is correct?

    TASK C.2

    A. A only
    B. B only
    C. Both A and B
    D. Neither A nor B

    Answer A is incorrect. Technician B is also correct.

    Answer B is incorrect. Technician A is also correct.

    Answer C is correct. Both Technicians are correct. You connect the battery positive cable first when hooking up a battery. One always disconnects the ground cable first when disconnecting a battery.

    Answer D is incorrect. Both Technicians are correct.

TASK D.1.11

5. What should a technician do when testing the operation of the trailer control valve?
   A. Move the handle to the fully applied position and record the air pressure.
   B. Drain the air system to 0 psi, and then allow pressure to build up again.
   C. Record the gauge reading of the air pressure first, and then drain the system.
   D. Listen for air leakage around the handle as the handle is moved to the applied position.

Answer A is correct. A technician should move the handle to the fully applied position and record the air pressure.

Answer B is incorrect. The air system does not have to be drained to test the trailer control valve.

Answer C is incorrect. As long as the trailer brakes are fully charged, the initial gauge reading is unnecessary.

Answer D is incorrect. Air leakage around the control handle should not occur.

2013 © Delmar, Cengage Learning

TASK D.2.10

6. What is the component shown in the illustration above?
   A. A one-way clutch
   B. A vertical yoke support
   C. An exhaust pipe support
   D. A center bearing support

Answer A is incorrect. A one-way clutch is not used in a driveline.

Answer B is incorrect. There is no such part as a vertical yoke support.

Answer C is incorrect. The exhaust pipe supports do not have that appearance.

Answer D is correct. A center support bearing is shown. Center support bearings are used when the distance between the transmission and the rear axle is too great to span with a single driveshaft. The center bearing is fastened to the frame and aligns the two connecting driveshafts. It consists of a stamped steel bracket used to align and fasten the bearing to the frame. A rubber mount inside the bracket surrounds the bearing, which supports the driveshaft.

TASK A.11

7. When checking the emission control systems of a gasoline-fueled truck, always test:
   A. The carburetor fuel inlet line.
   B. The positive crankcase ventilation (PCV) valve.
   C. The air cleaner vacuum hoses.
   D. The intake manifold service port.

Answer A is incorrect. The fuel inlet line is not part of an emissions systems check.

Answer B is correct. You should always check the PCV valve when any inspection of the emission control system is performed. The PCV system requires service at regular intervals to ensure that the system operates properly. Sludge and carbon can plug the PCV valve. Service the PCV valve by inspecting it, making sure that the valve is not clogged and it moves freely. Inspect all PCV system hoses for cracks and deterioration. Regular maintenance should include cleaning the PCV valve with a carburetor or fuel injector cleaner.

Answer C is incorrect. The air cleaner vacuum hoses are not part of an emissions systems check.

Answer D is incorrect. The intake manifold service port is not part of an emissions systems check.

8. A roadside inspection is performed while it is raining. All of the following defects would be reason to tag the vehicle out of service EXCEPT:

    A. The wiper blade on the driver's side is cracked and badly worn.

    B. The wiper motor is inoperative.

    C. The passenger's side wiper blade is cracked and badly worn.

    D. The driver's side wiper arm is distorted so the blade does not contact the windshield properly.

TASK B.12

Answer A is incorrect. When the wiper blade on the driver's side is cracked and badly worn, the vehicle should be tagged out of service.

Answer B is incorrect. When the wiper motor is inoperative, the vehicle should be tagged out of service.

Answer C is correct. When the passenger's side wiper blade is cracked and badly worn, the vehicle should not be tagged out of service.

Answer D is incorrect. If the driver's side wiper arm is distorted so the blade does not contact the windshield properly, the vehicle should be tagged out of service.

2013 © Delmar, Cengage Learning

9. In the illustration above, which test is being performed?

    A. A battery drain test

    B. A battery capacity test

    C. A battery open circuit voltage test

    D. A starter current draw test

TASK C.4

Answer A is incorrect. The test shown in the illustration is a battery capacity test.

Answer B is correct. The test shown in the illustration is a battery capacity test.

Answer C is incorrect. The test shown in the illustration is a battery capacity test.

Answer D is incorrect. The test shown in the illustration is a battery capacity test.

TASK D.4.1

10. Technician A says that multiple leaf springs should be replaced as an assembly. Technician B says that one end of semi-elliptic springs are free to move forward and backward. Who is correct?

    A. A only
    B. B only
    C. Both A and B
    D. Neither A nor B

    Answer A is incorrect. Technician B is also correct.

    Answer B is incorrect. Technician A is also correct.

    Answer C is correct. Both Technicians are correct. Semi-elliptical springs are usually mounted with a pin and bushing to the spring hanger and the other end is free to move forward and backward to compensate for changes in length. Multiple leaf springs should be replaced as an assembly.

    Answer D is incorrect. Both Technicians are correct.

TASK A.16

11. Technician A says that the air cleaner element should be replaced when 25 in. Hg air restriction is indicated. Technician B says that regardless of regulations, a truck must be serviced in the most cost-effective manner to the owner. Who is correct?

    A. A only
    B. B only
    C. Both A and B
    D. Neither A nor B

    Answer A is correct. Only Technician A is correct. When 25 in. Hg air restriction is reached, the element should be charged.

    Answer B is incorrect. Federal Motor Vehicle Safety Standards are very clear on confirming that a vehicle that is used commercially or industrially must be in a safe operating condition at all times, regardless of the cost to keep a vehicle in such a condition.

    Answer C is incorrect. Only Technician A is correct.

    Answer D is incorrect. Technician A is correct.

TASK C.11

12. When aiming headlights with a headlight aligning screen:

    A. The vehicle does not need to be parallel to the aligning screen.
    B. The tire inflation pressure does not affect headlight alignment.
    C. The headlight alignment may be changed by installing new sealed beams.
    D. The headlight alignment is changed by turning adjusting screws on the headlight assembly.

    Answer A is incorrect. The vehicle does need to be parallel to the aligning screen.

    Answer B is incorrect. The tire inflation pressure does affect headlight alignment.

    Answer C is incorrect. The headlight alignment is not changed by installing new sealed beams.

    Answer D is correct. The headlight alignment is changed by turning adjusting screws on the headlight assembly.

13. Technician A says that technicians who perform mobile refrigerant system service must be certified. Technician B always makes sure both hand valves on the manifold gauge set are closed before connecting the hoses to the vehicle. Who is correct?

TASKS B.28, B.29, B.30

   A. A only
   B. B only
   C. Both A and B
   D. Neither A nor B

Answer A is incorrect. Technician B is also correct.

Answer B is incorrect. Technician A is also correct.

Answer C is correct. Both Technicians are correct. Technicians performing refrigeration service must be certified to do so. A manifold gauge set is used to recycle, recharge, and diagnose A/C systems. The manifold gauge set is designed to access the refrigeration circuit and control refrigerant flow when adding or removing refrigerant. When the manifold test set is connected into the system, pressure is registered on both of the gauges at all times. Make certain that both hand valves on the manifold gauge set are closed before connecting to the system to avoid releasing refrigerant or contaminating the system.

Answer D is incorrect. Both Technicians are correct.

14. How should a truck's engine mounts be dynamically checked?

   A. Visually inspecting for cracks and/or deterioration
   B. Correlating information from the driver and the vehicle's maintenance history
   C. Observing the engine mounts while the engine is torqued with its brakes applied
   D. Automatically deadlining a vehicle due to the driver's suspicions

TASK A.6

Answer A is incorrect. A simple visual inspection without having the engine assembly "moving" would be inaccurate since the mounts would not be under any dynamic stress.

Answer B is incorrect. Correlating information from the driver and from the vehicle's maintenance history can only confirm a proper "dynamic" test of a truck's engine mounts.

Answer C is correct. The correct way to check engine mounts is to observe the engine and its mounts while the engine is being torqued with its brakes applied.

Answer D is incorrect. A vehicle should never be deadlined because of a driver's suspicion.

15. While checking fan blades, Technician A says a cracked fan blade may cause a clicking or creaking noise with the engine idling. Technician B says you should not stand or position your body in line with the fan while increasing the engine speed. Who is correct?

TASK A.18

   A. A only
   B. B only
   C. Both A and B
   D. Neither A nor B

Answer A is incorrect. Technician B is also correct.

Answer B is incorrect. Technician A is also correct.

Answer C is correct. Both Technicians are correct. A cracked fan blade may cause a clicking or creaking noise with the engine idling. You should not stand or position your body in line with the fan while increasing the engine speed.

Answer D is incorrect. Both Technicians are correct.

TASK B.26

16. How would suspension height of the illustration above most likely be checked?

A. Measure from the road surface to the lower plate.

B. Measure between the upper and lower plates.

C. Measure from the road surface to the upper plate.

D. Measure the angle of the regulator valve arm.

Answer A is incorrect. Measuring from the road surface to the lower plate would be inaccurate.

Answer B is correct. A cab air suspension system is shown in the illustration. When checking cab ride height, the truck should be on a level surface and the cab should appear to be level. Make sure system air pressure is at specified pressure before performing measurements. Use a tape measure to check cab ride height at the location specified by the manufacturer. This is usually between the upper and lower plates. Compare readings with those specified by the manufacturer and adjust the leveling valve as necessary.

Answer C is incorrect. Measuring from the road surface to the upper plate would be inaccurate.

Answer D is incorrect. Measuring the angle of the regulator valve arm would be inaccurate.

TASK D.1.13

17. Technician A says a 45-degree elbow in an air brake line may be replaced with a 90-degree elbow without affecting brake operation. Technician B says that if a trailer service system has an air fitting that is smaller than the original line; the air timing/balance is affected. Who is correct?

A. A only

B. B only

C. Both A and B

D. Neither A nor B

Answer A is incorrect. If a 45-degree elbow in an air brake line is replaced with a 90-degree elbow, brake timing is adversely affected.

Answer B is correct. Only Technician B is correct. A smaller than specified air line slows the brake application.

Answer C is incorrect. Only Technician B is correct.

Answer D is incorrect. Technician B is correct.

18. Which of the following does the push hydraulic circuit in a cab lift system do?

    A. Lowers the cab back from a fully extended position

    B. Raises the cab from fully extended to the 45-degree tilt position

    C. Allows the cab to descend to the 60-degree overhead position

    D. Raises the cab from the lowered position to the desired tilt location

TASK B.24

Answer A is incorrect. The pull circuit allows the cab to come back over the engine.

Answer B is incorrect. The pull circuit allows the cab to come to the 45-degree tilt position.

Answer C is incorrect. The pull circuit allows the cab to descend to the 60-degree overhead position.

Answer D is correct. The push circuit raises the cab from the lowered position to the desired tilt position. Remember that in most systems, whenever raising or lowering the cab, stop working the hydraulic pump once the cab goes over center. The cab falls at a controlled rate and continued pumping could lock up the tilt cylinder. Inspect the hydraulic cylinders for external leakage and determine if the cab can be raised and lowered. Check for internal leakage by observing bleed down. Make sure the mechanical stops engage and hold safely.

19. When adjusting a clutch linkage, Technician A says that the pedal free travel should be about 1½ to 2 inches (38.1 to 50.8 mm). Technician B says that release bearing to clutch brake travel should be less than ½ inch (12.7 mm). Who is correct?

    A. A only

    B. B only

    C. Both A and B

    D. Neither A nor B

TASK D.2.4

Answer A is correct. Only Technician A is correct. The pedal free travel should be about 1½ to 2 inches (38.1 to 50.8 mm). This clearance ensures full clutch engagement, and no interference from the linkage should occur.

Answer B is incorrect. If the release bearing to clutch brake travel is less than ½ inch the release bearing will contact the clutch brake too soon, causing clutch brake damage, and complete clutch disengagement may not be possible.

Answer C is incorrect. Only Technician A is correct.

Answer D is incorrect. Technician A is correct.

20. Technician A says an ohmmeter should be used to check for a short circuit between two circuits. Technician B says the battery must be fully charged before checking a circuit for current draw. Who is correct?

    A. A only

    B. B only

    C. Both A and B

    D. Neither A nor B

TASK C.10

Answer A is incorrect. An ohmmeter is not used to check for a short circuit between two circuits.

Answer B is correct. Only Technician B is correct. The battery must be fully charged before checking a circuit for current draw.

Answer C is incorrect. Only Technician B is correct.

Answer D is incorrect. Technician B is correct.

TASKS D.3.14,
D.3.18, D.4.3,
D.4.1

21. What is the LEAST LIKELY cause of a vehicle wandering?

    A. Mismatched or under-inflated tires

    B. Unmatched spring design/load capacity spring assemblies

    C. One or more broken spring leaves

    D. Frame alignment within manufacturer's specifications

    Answer A is incorrect. Mismatched or under-inflated tires will cause vehicle wandering.

    Answer B is incorrect. Springs matched with different spring rates will cause an imbalance, sagging, and/or loss of designed load-carrying ability, and could also possibly affect the designed riding geometry of the vehicle and thus cause wandering.

    Answer C is incorrect. Broken or missing spring leaves will cause a reduction in the load-carrying ability and of overall spring rate of the affected vehicle and will cause wandering.

    Answer D is correct. A frame has to be within manufacturer's specifications.

TASKS D.1.21,
D.1.24

22. Technician A says that when checking the condition of hydraulic drum brakes on a vehicle, to look for leakage and contamination around the wheel cylinders. Technician B inspects the brake lines for evidence of corrosion. Who is correct?

    A. A only

    B. B only

    C. Both A and B

    D. Neither A nor B

    Answer A is incorrect. Technician B is also correct.

    Answer B is incorrect. Technician A is also correct.

    Answer C is correct. Both Technicians are correct. A proper inspection of a heavy-duty truck hydraulic braking system should include a visual of wheel cylinders. Look for evidence of leaks on the inboard side of the tire and wheel. Leakage and contamination around the wheel cylinders generally indicates leakage and that repairs are needed. Leaking wheel cylinders must be reconditioned or replaced. Brake lines should be in good condition with no corrosion or damage that may compromise system performance. Leaking or damaged brake lines must be replaced.

    Answer D is incorrect. Both Technicians are correct.

TASK C.10

23. While checking fuses in a tractor/trailer, the technician finds an open fuse. What is the next step?

    A. Replace the fuse with the next higher-amperage fuse.

    B. Check the affected circuit for a short to ground.

    C. Check the affected circuit for an open.

    D. Install a circuit breaker of the same amperage as the fuse.

    Answer A is incorrect. The technician should not replace the fuse with the next higher-amperage fuse.

    Answer B is correct. The technician should check the affected circuit for a short to ground.

    Answer C is incorrect. An open circuit does not cause a blown fuse.

    Answer D is incorrect. The technician should not install a circuit breaker of the same amperage as the fuse.

24. The Federal Highway Administration has set up a minimum inspection standards program. Technician A says that if a vehicle decal is used, a copy of the inspection form must be kept on file and should indicate where an inspector can get a copy of it. Technician B says that a tractor/trailer is considered to be one vehicle, requiring only one decal. Who is correct?

TASK A.1

    A. A only
    B. B only
    C. Both A and B
    D. Neither A nor B

    Answer A is Correct. Each vehicle must carry proof that an inspection was completed. Proof can either be a copy of the inspection form kept on the vehicle or a decal. If using a decal, a copy of the inspection form must be kept on file.

    Answer B is incorrect. A tractor/trailer is considered to be two vehicles, each requiring a separate decal. A converter dolly is also a separate vehicle.

    Answer C is incorrect. Only Technician A is correct.

    Answer D is incorrect. Technician A is correct.

25. Most vehicle manufacturers recommend checking the level of power steering fluid at which working temperature?

    A. 100°F
    B. 140°F

TASK D.3.2

    C. 90°F
    D. 175°F

    Answer A is incorrect. 100°F is too low.

    Answer B is incorrect. 140°F is too low.

    Answer C is incorrect. 90°F is too low.

    Answer D is correct. 175°F is the correct value. Most OEM truck manufacturers recommend checking the power steering fluid level at an operating or working temperature of 175°F (79°C). With the engine at 1,000 rpm or less, turn the steering wheel slowly and completely in each direction several times to raise the fluid temperature. Check the reservoir for foaming as a sign of aerated fluid. The fluid level in the reservoir should be at the hot full mark on the dipstick.

26. A high resistance is suspected in the starting motor ground circuit on a negative ground electrical system. Technician A says to connect the voltmeter leads from the starting motor housing to the positive battery terminal. Technician B says the fuel injection system should be disabled and the engine cranked to read the voltage drop. Who is correct?

TASK C.5

    A. A only
    B. B only
    C. Both A and B
    D. Neither A nor B

    Answer A is incorrect. The voltmeter leads should be connected from the starting motor housing to the negative battery terminal.

    Answer B is correct. Only Technician B is correct. The fuel injection system should be disabled and the engine cranked to read the voltage drop.

    Answer C is incorrect. Only Technician B is correct.

    Answer D is incorrect. Technician B is correct.

TASK D.2.1

27. Technician A says that a clutch brake needs to be inspected for wear and fatigue. Technician B says that providing the clutch brake drive lugs are intact, it will always function properly. Who is correct?

    A. A only

    B. B only

    C. Both A and B

    D. Neither A nor B

Answer A is correct. Only Technician A is correct. A clutch brake needs to be inspected for wear and fatigue. Worn clutch brake friction surfaces can lead to clashing (grinding) when moving the gear lever into reverse or low gear while the vehicle is stationary.

Answer B is incorrect. The friction face needs to be inspected. The clutch brake is a circular disc with a friction surface that is mounted on the transmission input spline shaft using two tangs that engage machined slots. The clutch brake is mounted between the release bearing and the transmission. Its purpose is to slow or stop the transmission input shaft from rotating in order to allow gears to be engaged without clashing and to keep transmission gear damage to a minimum. Clutch brakes are used only on vehicles with non-synchronized transmission.

Answer C is incorrect. Only Technician A is correct.

Answer D is incorrect. Technician A is correct.

28. Wires for fusible links should be:

    A. The same gauge size as the circuit wire.

    B. The same color as the circuit wire.

    C. Two gauge sizes larger than the circuit wire.

    D. A smaller gauge size than the circuit wire.

TASK C.7

Answer A is incorrect. Wires for fusible links should be smaller than the circuit wire.

Answer B is incorrect. Wires for fusible links should be smaller than the circuit wire.

Answer C is incorrect. Wires for fusible links should be smaller than the circuit wire.

Answer D is correct. Wires for fusible links should be smaller than the circuit wire.

**TASKS D.2.14,
D.2.15**

29. All of the following could cause a drive axle failure of a single drive axle vehicle EXCEPT:

    A. Incorrect ratio.

    B. Normal wear.

    C. Lubrication failure.

    D. Fatigue.

Answer A is correct. An incorrect ratio will only affect the truck's mileage and pulling ability, not cause failure.

Answer B is incorrect. Normal wear always eventually causes a failure.

Answer C is incorrect. Incorrect or lack of lubrication will always cause a failure.

Answer D is incorrect. Fatigue will cause a drive axle failure.

30. A cooling system is being pressure tested. Technician A says that a radiator pressure cap reading of 15 psi is normal. Technician B says that it is necessary to warm the engine to normal operating temperature before conducting a pressure test. Who is correct?

TASK A.20

    A.   A only

    B.   B only

    C.   Both A and B

    D.   Neither A nor B

Answer A is correct. Only Technician A is correct. A cooling system pressure reading of 15 psi is normal for pressure testing.

Answer B is incorrect. A cooling system pressure tester does not generally require a normal engine operating temperature for operation. A cooling system pressure tester has a manually operated pump to raise system pressure without raising its temperature.

Answer C is incorrect. Only Technician A is correct.

Answer D is incorrect. Technician A is correct.

31. The headlight switch has failed in the circuit shown in the illustration above, Technician A says the parking lights are not affected. Technician B says the stoplights will fail to operate. Who is correct?

TASK C.11

    A.   A only

    B.   B only

    C.   Both A and B

    D.   Neither A nor B

Answer A is incorrect. The parking lights will be affected.

Answer B is incorrect. The stoplights will continue to operate because voltage supply to these lights is independent of the headlight switch.

Answer C is incorrect. Neither Technician is correct.

Answer D is correct. Neither Technician is correct. The parking lights are part of the headlight circuit and work in conjunction with the headlight switch, as a result, failure of the switch will affect operation of the parking lights. The stoplights operate separately and will continue to operate.

TASK D.1.27

32. Technician A says that trucks with hydraulic-assist power brakes are not required to have a back-up system. Technician B says that on systems that do incorporate a back-up that the system should be checked during each inspection. Who is correct?

  A. A only
  B. B only
  C. Both A and B
  D. Neither A nor B

Answer A is incorrect. Trucks equipped with hydraulic-assist power brakes require back-up to supplement power braking in the event the engine stops while the truck is moving.

Answer B is correct. Hydraulic assist back-up systems and associated warning devices should be checked during each vehicle inspection.

Answer C is incorrect. Only Technician B is correct.

Answer D is incorrect. Technician B is correct.

TASKS
E.1, E.5

33. Technician A says that checking clutch operation and transmission shifting is part of a drive test during a preventive maintenance inspection (PMI). Technician B claims that when a drive test is performed, road speed governor actual speed should be checked. Who is correct?

  A. A only
  B. B only
  C. Both A and B
  D. Neither A nor B

Answer A is incorrect. Technician B is also correct.

Answer B is incorrect. Technician A is also correct.

Answer C is correct. Both Technicians are correct. Checking the operation of the clutch and gearshift is part of a drive test during a PMI. When a drive test is performed, the speed at which the governor is set is always checked.

Answer D is incorrect. Both Technicians are correct.

TASK D.1.9

34. The tractor protection valve closes to protect the tractor air supply if the air pressure in the supply line drops below what psi value?

  A. 20 psi (138 kPa)
  B. 55 psi (379 kPa)
  C. 75 psi (517 kPa)
  D. 80 psi (552 kPa)

Answer A is correct. The tractor protection valve closes to protect the tractor air supply if the supply line pressure drops below 20 psi (138 kPa).

Answer B is incorrect. The tractor protection valve closes to protect the tractor air supply if the supply line pressure drops below 20 psi (138 kPa).

Answer C is incorrect. The tractor protection valve closes to protect the tractor air supply if the supply line pressure drops below 20 psi (138 kPa).

Answer D is incorrect. The tractor protection valve closes to protect the tractor air supply if the supply line pressure drops below 20 psi (138 kPa).

35. A heavy-duty truck is found to have an exhaust leak. Technician A intends to either repair it immediately or deadline the truck. Technician B says that the vehicle can only be put out of service if the leaking exhaust fumes enter the driver/sleeper compartment or if they are likely to result in charring, burning, or damaging of the wiring, fuel supply, or other combustible parts. Who is correct?

TASK A.13

   A.   A only
   B.   B only
   C.   Both A and B
   D.   Neither A nor B

   Answer A is incorrect. A truck may not necessarily be pulled out of service if an exhaust leak does not meet the CVSA out-of-service criteria.

   Answer B is correct. Only Technician B is correct. The vehicle can only be put out of service if the leaking exhaust fumes enter the driver/sleeper compartment or if they're likely to result in charring, burning, or damaging of the wiring, fuel supply or other combustible parts.

   Answer C is incorrect. Only Technician B is correct.

   Answer D is incorrect. Technician B is correct.

36. An applied pressure drop test is being performed on a straight truck air brake system. Technician A says that the pressure drop should be no more than 6 psi in two minutes. Technician B says you hold down the brake pedal for the first minute, and then release it. Who is correct?

TASKS D.1.4, D.1.5

   A.   A only
   B.   B only
   C.   Both A and B
   D.   Neither A nor B

   Answer A is incorrect. The applied pressure drop on a straight truck with no trailer should be no more than 4 psi in two minutes.

   Answer B is incorrect. The technician holds the brake pedal down for the entire test to ensure a proper reading.

   Answer C is incorrect. Neither Technician is correct.

   Answer D is correct. Neither Technician is correct. Before performing a pressure-drop test, the system must be at full pressure. Make and maintain a brake application. Allow pressure to stabilize for one minute, and then begin timing for two minutes while watching the dash gauge. A block of wood may be used to hold the brake pedal down during the pressure-drop test.

37. When discussing auxiliary power unit (APU) maintenance, all of the following are true EXCEPT:

TASK B.32

   A.   Battery-powered systems are more maintenance-intensive than engine-based systems.
   B.   On engine-based systems, specific points such as the radiator and condenser fins should be cleaned and/or inspected.
   C.   On battery-powered systems, a visual inspection should be performed as well as checking batteries and connections.
   D.   Many APU failures can be traced to environmental factors such as vibration, corrosive road spray, and extreme heat and cold.

   Answer A is correct. A battery-powered system is relatively trouble free and generally requires less maintenance than engine-based systems.

   Answer B is incorrect. Specific points on the engine-based system should be cleaned and/or inspected on regular service maintenance intervals.

   Answer C is incorrect. On battery-powered systems, a visual inspection should be performed as well as checking the condition of the battery, battery mounts, cables for corrosion and fraying, and checking connections.

   Answer D is incorrect. The APU is located on the truck frame and exposed to many environmental factors. Corrosive road spray, extreme heat and cold, road vibrations, and foreign objects can lead to APU failures.

TASK A.28

38. Technician A says an engine oil sample should be taken during every PMI. Technician B says to take an engine oil sample mid-way through the sump drain-off. Who is correct?
    A. A only
    B. B only
    C. Both A and B
    D. Neither A nor B

Answer A is incorrect. Technician B is also correct.

Answer B is incorrect. Technician A is also correct.

Answer C is correct. Both Technicians are correct. During a normal PMI, an engine oil sample is taken for inspection of its overall lubricating qualities, plus metal and dirt content. This oil may also be tested for corrosion-causing agents such as sulfuric and hydrochloric acids, and some esters. It is for this reason that the engine oil should be at normal operating temperature to ensure that all possible contaminants and corrosives be contained in the oil in its normal operating condition. The oil sample should be taken during an oil change at the midpoint of the drainage. The test bottle should be clean before the sample, and sealed and labeled afterward.

Answer D is incorrect. Both Technicians are correct.

TASKS D.4.9, D.4.15

39. Technician A says that a circle inspection includes items such as torn mud flaps, leaking hoses, and outdated permits. Technician B says to perform a circle inspection first before any other inspection as part of a PMI. Who is correct?
    A. A only
    B. B only
    C. Both A and B
    D. Neither A nor B

Answer A is incorrect. Technician B is also correct.

Answer B is incorrect. Technician A is also correct.

Answer C is correct. Both Technicians are correct. A circle inspection includes items such as torn mud flaps, leaking hoses, and outdated permits. Performing a circle inspection first before any other inspection is part of a PMI.

Answer D is incorrect. Both Technicians are correct.

TASK D.1.25

40. A vehicle with vacuum-boosted brakes is tagged out of service if in the vacuum system the:
    A. Vacuum never exceeds 15 in. Hg.
    B. Vacuum lines appear to be OK, but the engine lacks performance.
    C. Vacuum reserve is insufficient to permit one full brake application after the engine is shut off.
    D. System vacuum bleeds down over 50 percent after engine shut off within two minutes.

Answer A is incorrect. The low system vacuum does not necessarily condemn a vehicle to be out of service, but will require a thorough diagnostic inspection by a certified technician depending upon the severity of the loss.

Answer B is incorrect. An engine with low performance can have many causes, not exclusively low vacuum. Outside of direct leakage to the atmosphere, low vacuum is generally a result of low engine performance, not a cause of it.

Answer C is correct. The vacuum reserve is insufficient to permit one full brake application after the engine is shut off.

Answer D is incorrect. There are no set system-wide vacuum bleed-down standards.

41. Technician A says that slack adjusters should be lubricated when brakes are inspected. Technician B says that only the rear brake chamber pushrods have to be checked for stress cracks. Who is correct?

    TASK D.1.14

    A. A only

    B. B only

    C. Both A and B

    D. Neither A nor B

    Answer A is correct. Only Technician A is correct. You always lubricate the slack adjusters whenever an inspection of the brakes is scheduled. Slack adjusters are lubricated by grease or automatic lubing systems. The seals in slack adjusters are always installed with the lip angle facing outward. When grease is pumped into the slack adjuster, grease exits easily past the seal lip when the internal lubrication circuit has been charged.

    Answer B is incorrect. All brake chamber pushrods must be checked during a brake inspection. Air brake chambers should be checked for signs of damage. Visually check for damage from road hazards and look for cracks around mounting studs. Check the air chamber bracket for cracks, looseness, or audible air leaks at the brake chamber that might indicate a ruptured diaphragm, loose chamber clamp, or other failure.

    Answer C is incorrect. Only Technician A is correct.

    Answer D is incorrect. Technician A is correct.

42. When adjusting wheel bearings, Technician A says that TMC wheel-end procedure should be followed at all times. Technician B says that wheel bearings set with preload are likely to fail prematurely. Who is correct?

    TASK D.3.9

    A. A only

    B. B only

    C. Both A and B

    D. Neither A nor B

    Answer A is incorrect. Technician B is also correct.

    Answer B is incorrect. Technician A is also correct.

    Answer C is correct. Both Technicians are correct. Maladjusted bearings fail rapidly, especially when preloaded. The result of high preload on a bearing can be to friction-weld it to the axle. All bearing manufacturers in North America have endorsed the Truck Maintenance Council (TMC) method of adjusting tapered roller wheel bearings. The bearing end-play must be measured with a dial indicator. The required specification is between 0.001 and 0.005 inch. End-play must be present. If at least 0.001 inch end-play is not present, the adjustment procedure must be repeated.

    Answer D is incorrect. Both Technicians are correct.

TASK A.8

43. Technician A says any type of wire may be used to replace a burned fusible link as long as the gauge is one size smaller than the wire in the circuit being protected. Technician B says that a circuit breaker should be replaced after repairing a short in a circuit. Who is correct?

A. A only

B. B only

C. Both A and B

D. Neither A nor B

Answer A is incorrect. A burned fusible link must be replaced with a length of specified fusible link wire.

Answer B is incorrect. Circuit breakers can be reset.

Answer C is incorrect. Neither Technician is correct.

Answer D is correct. Neither Technician is correct. A burned fusible link must be replaced with a length of the specified fusible link wire. A circuit breaker resets itself or can be reset and does not have to be replaced after repairing a short in a circuit.

TASK D.4.2

44. After a stretch of rough road is driven, the steering wheel continues to shake for a few seconds. Which of these could be the cause?

A. Leaking front shock absorbers

B. Low tire pressure

C. Rusted rear shock absorbers

D. Missing jounce bumper

Answer A is correct. Leaking, worn-out front shock absorbers can cause the steering wheel to shake for a few seconds after a stretch of rough road.

Answer B is incorrect. Low tire pressure would not cause wheel shake.

Answer C is incorrect. Rusted rear shock absorbers would not cause wheel shake.

Answer D is incorrect. A missing jounce bumper would cause a bang on a rough road but not a shake.

TASK D.4.7

45. Which of the following is the standard fifth-wheel plate height?

A. 39 inches

B. 43 inches

C. 47 inches

D. 49 inches

Answer A is incorrect. 39 inches is not considered a standard fifth-wheel height.

Answer B is incorrect. 43 inches is not considered a standard fifth-wheel height.

Answer C is correct. 47 inches is considered the standard fifth-wheel height. Most trailers are designed with an intended 47 inches upper-coupler (kingpin) plate height. This, plus the fact that any trailers operated on North American roads cannot exceed a height of 13 feet, 6 inches, determines the plate height of fifth wheels.

Answer D is incorrect. 49 inches is not considered a standard fifth-wheel height.

46. A maintenance-free battery is low on electrolyte and the built-in hydrometer indicates yellow. Technician A says a faulty voltage regulator may cause this problem. Technician B says high resistance in the circuit between the alternator battery terminal and the positive battery terminal may be the cause. Who is correct?

TASK C.3

A. A only

B. B only

C. Both A and B

D. Neither A nor B

Answer A is correct. Only Technician A is correct. A faulty voltage regulator may cause overcharging and low battery electrolyte.

Answer B is incorrect. High resistance in the circuit between the alternator battery terminal and the positive battery terminal causes undercharging of the battery.

Answer C is incorrect. Only Technician A is correct.

Answer D is incorrect. Technician A is correct.

47. On an automatic transmission, the following methods can be used for checking transmission shift points EXCEPT:

TASK E.4

A. Stall test method.

B. Speedometer readings.

C. Test stand tailshaft speed.

D. Engine RPM.

Answer A is correct. The stall test is a separate test used for testing other types of transmission problems, not shift points.

Answer B is incorrect. Using speedometer readings is an acceptable troubleshooting method for checking shift point irregularities and to determine if adjustments are needed.

Answer C is incorrect. Using test stand tailshaft speed method is the most accurate testing method. This test is performed using a valve body and governor test stand.

Answer D is incorrect. Using RPM readings is an acceptable troubleshooting method. Using this test or the speedometer test require using a chassis dynamometer or taking a test drive.

48. A highway tractor is having a full PMI performed. Technician A says that outboard S-cam bushings should be pressure-lubricated until grease flows from the outboard seal. Technician B says to use the maximum pressure available from pressure lubricating equipment to aid in flushing contaminants from lubricated components. Who is correct?

TASK D.1.19

A. A only

B. B only

C. Both A and B

D. Neither A nor B

Answer A is incorrect. The outboard S-cam bushings should be lubricated until grease appears at the inboard seal; if grease exits through the outboard seal it will end up in the drum assembly.

Answer B is incorrect. Low-pressure grease is sufficient to lubricate most chassis components and high-pressure grease can cause seal damage and its use should be avoided except for lube operations such as spring pins.

Answer C is incorrect. Neither Technician is correct.

Answer D is correct. Neither Technician is correct. Failure of the outboard seal will allow grease to enter the drum assembly and damage the drum and brake linings. Pressure lubricating equipment should be checked for proper operation and adjustment before use.

TASK A.22

49. An excessively high coolant level in the recovery reservoir may be caused by any of these problems EXCEPT:

   A. Restricted radiator tubes.

   B. A thermostat that is stuck open.

   C. A blown head gasket.

   D. An inoperative air-operated cooling fan clutch.

   Answer A is incorrect. Restricted radiator tubes may cause excessive coolant level in the recovery reservoir.

   Answer B is correct. A thermostat that is stuck open does not cause excessive coolant level in the recovery reservoir.

   Answer C is incorrect. A blown head gasket could allow combustion chamber gases into the coolant, which would raise the coolant level in the recovery reservoir.

   Answer D is incorrect. An inoperative air-operated cooling fan clutch may cause excessive coolant level in the recovery reservoir.

TASKS D.3.15, D.3.17

50. Wheel assemblies are being discussed. Technician A says hub-piloted disc wheels are less susceptible to loosening and cracking than stud-piloted wheels. Technician B says that disc wheels generally experience more alignment and balance problems than do cast spoke designs. Who is correct?

   A. A only

   B. B only

   C. Both A and B

   D. Neither A nor B

   Answer A is correct. Only Technician A is correct. Hub-piloted wheels are less susceptible to loosening and cracking than stud-piloted wheels. The hub-piloted system simplifies centering and clamping wheels to the hubs because the hub centers the wheel and the nuts and studs provide only clamping force. The hub-piloted wheel uses one cap nut and eliminates the need for inner cap nuts.

   Answer B is incorrect. Generally, spoke wheels experience greater alignment and balance problems than disc designs. Proper installation procedures are critical on spoke wheels. If the rim clamps are not properly installed, wheel out-of-round or wobble may be excessive.

   Answer C is incorrect. Only Technician A is correct.

   Answer D is incorrect. Technician A is correct.

PREPARATION EXAM 3—ANSWER KEY

1.	A	18.	A	35.	C
2.	A	19.	A	36.	C
3.	C	20.	D	37.	D
4.	A	21.	A	38.	A
5.	A	22.	B	39.	B
6.	C	23.	B	40.	B
7.	B	24.	A	41.	C
8.	C	25.	C	42.	D
9.	C	26.	D	43.	A
10.	B	27.	D	44.	B
11.	B	28.	C	45.	A
12.	D	29.	B	46.	B
13.	D	30.	C	47.	A
14.	D	31.	D	48.	A
15.	C	32.	D	49.	B
16.	D	33.	D	50.	A
17.	D	34.	A		

PREPARATION EXAM 3—EXPLANATIONS

1. Technician A says that all current tractor-trailers are required to have automatic slack adjusters. Technician B says that automatic slack adjusters increase the legal free stroke dimension. Who is correct?

    TASK D.1.17

    A. A only

    B. B only

    C. Both A and B

    D. Neither A nor B

    Answer A is correct. Only Technician A is correct. All current (manufactured after 1998) tractors and trailers are required to have automatic slack adjusters.

    Answer B is incorrect. Automatic slack adjusters do not increase the allowable free stroke dimension.

    Answer C is incorrect. Only Technician A is correct.

    Answer D is incorrect. Technician A is correct.

TASK C.8

2. An alternator with a 90-ampere rating produces 45 amperes during an output test. What is the most likely cause of this problem?

A. An alternator drive belt that is bottomed in the pulley

B. An open circuit in the rotor field coil

C. A misaligned alternator pulley

D. A broken alternator brush lead

Answer A is correct. An alternator drive belt that is bottomed in the pulley causes belt slipping and low output.

Answer B is incorrect. An open circuit in the rotor field coil causes zero output.

Answer C is incorrect. A misaligned alternator pulley does not affect output.

Answer D is incorrect. A broken alternator brush lead causes zero output.

TASK A.13

3. All of these conditions could be Commercial Vehicle Safety Alliance (CVSA) out-of-service criteria EXCEPT:

A. An exhaust leak under the sleeper.

B. A wiring harness that contacts the exhaust.

C. An inaccurate exhaust temperature gauge.

D. A missing exhaust heat shield.

Answer A is incorrect. An exhaust leak under the cab or sleeper is a CVSA out-of-service condition.

Answer B is incorrect. Incorrect routing of a wiring harness, air lines, or fuel lines allowing them to contact the exhaust are CVSA out-of-service conditions.

Answer C is correct. An inaccurate exhaust temperature gauge would not by itself place a truck out of service.

Answer D is incorrect. Missing heat shields and guards could deadline a truck.

TASK D.1.17

4. A slack adjuster is being adjusted. Technician A says wear between the clevis pin and yoke will cause the stroke to be too long. Technician B says that an automatic slack adjuster will need frequent adjustment. Who is correct?

A. A only

B. B only

C. Both A and B

D. Neither A nor B

Answer A is correct. Only Technician A is correct. Wear between the clevis pin and yoke causes mechanical slop, which results in overstroking.

Answer B is incorrect. Properly functioning automatic slack adjusters should not require frequent adjustment. Manually adjusting automatic slack adjusters is dangerous and should not be done, except during installation or in an emergency to move the vehicle to a repair facility, because manual adjustment of this brake component (1) fails to address the true reason why the brakes are not maintaining adjustment, giving the operator a false sense of security about the effectiveness of the brakes, which are likely to go out of adjustment again soon, and (2) causes abnormal wear to the internal adjusting mechanism for most automatic slack adjusters, which may lead to failure of this brake component.

Answer C is incorrect. Only Technician A is correct.

Answer D is incorrect. Technician A is correct.

5. What should you do to eliminate a mildew smell from the air conditioner (A/C) output air?

TASK B.31

    A. Remove the evaporator case, clean it with a vinegar and water solution, and dry it thoroughly before reinstallation.

    B. Spray a disinfectant into the panel outlets.

    C. Pour a small amount of alcohol into the air intake plenum.

    D. Place an automotive deodorizer under the dash.

Answer A is correct. Vinegar will kill the mold and mildew that are the source of the odor.

Answer B is incorrect. This method will simply mask the odor and not eliminate the problem.

Answer C is incorrect. This method poses a possible fire hazard.

Answer D is incorrect. This method will simply mask the odor and not eliminate the problem.

6. An inversion valve is being tested on a heavy-duty truck. Technician A says to drain both the supply tank and the primary circuit reservoir before beginning the test. Technician B verifies that secondary circuit pressure is at least 100 psi before conducting the test. Who is correct?

TASK D.1.8

    A. A only

    B. B only

    C. Both A and B

    D. Neither A nor B

Answer A is incorrect. Technician B is also correct.

Answer B is incorrect. Technician A is also correct.

Answer C is correct. Both Technicians are correct. Both the supply tank and the primary circuit reservoir must be drained before beginning the test. The air system should be fully charged (minimum 100 psi), and the spring parking brakes must be released. Next, locate and drain the supply tank and then the primary axle service reservoir completely by means of the draincocks. The primary gauge needle will drop to 0 psi, indicating a pressure loss for that system, which is normal. The secondary gauge needle must remain at system pressure (minimum 100 psi) to start the check of the inversion valve. Depress the foot valve in a normal manner. The spring brakes on the rear axle(s) will apply as the air is exhausted from the brake chambers. Release the foot valve, and the spring brakes will release as air is supplied to them from the secondary reservoir. If the spring brakes do not or only partially release, then the inversion valve has failed and should be replaced. An inversion valve is a failsafe device and does not operate when brakes are functioning properly.

Answer D is incorrect. Both Technicians are correct.

2013 © Delmar, Cengage Learning

TASK C.2

7. What type of battery do the arrows in the illustration above indicate?

A.   Top terminal, low maintenance

B.   Stud terminal, maintenance free

C.   Side terminal, maintenance free

D.   Inverted terminal, high maintenance

Answer A is incorrect. A stud terminal maintenance-free battery is shown.

Answer B is correct. A stud terminal maintenance-free battery is shown. Top stud terminals incorporate a threaded terminal design. Side terminal batteries use internal threaded posts positioned in the side of the battery case that require a special bolt to connect the cables. Post terminal batteries use round, tapered terminals on top of the battery. The positive post is larger in diameter than the negative post. This helps to identify the polarity of the battery post and prevent connecting the battery in reverse polarity.

Answer C is incorrect. A stud terminal maintenance-free battery is shown.

Answer D is incorrect. A stud terminal maintenance-free battery is shown.

TASK A.19

8. Technician A says that a possible cause of an above-normal engine temperature is a faulty fan clutch. Technician B claims that plugged radiator fins may cause overheating. Who is correct?

A.   A only

B.   B only

C.   Both A and B

D.   Neither A nor B

Answer A is incorrect. Technician B is also correct.

Answer B is incorrect. Technician A is also correct.

Answer C is correct. Both Technicians are correct. A faulty fan clutch can cause above-normal engine temperatures or overheating. Fan clutch types may be: viscous, thermatic, air apply, air release, electric, and hydraulic. Each type has different characteristics and should first be identified to avoid improper diagnosis. Most pneumatic fan clutch failures are caused by air leaks. Viscous and hydraulic units should not exhibit any signs of fluid leakage. Radiators, because of their location, are susceptible to damage and restriction. Buildup of road debris and summer bugs can severely compromise the radiator's ability to effectively cool. Radiators should be cleaned externally using either a low-pressure steamer or regular hose, detergent, and a soft nylon bristle brush. Never use a high-pressure washer because this will almost certainly result in damaging the cooling fins.

Answer D is incorrect. Both Technicians are correct.

9. If the floating caliper on a hydraulic brake truck is seized, which of the following conditions will result?

TASK D.1.24

A. Brake pedal fade

B. Brake pedal pulsations

C. Reduced braking force

D. Brake grabbing

Answer A is incorrect. A seized floating caliper on a hydraulic brake system does not cause brake pedal fade.

Answer B is incorrect. A seized floating caliper on a hydraulic brake system does not cause brake pedal pulsations.

Answer C is correct. A seized floating caliper on a hydraulic brake system causes reduced braking force.

Answer D is incorrect. A seized floating caliper on a hydraulic brake system does not cause brake grabbing.

10. A windshield is being inspected during a routine inspection. Technician A says that the vehicle need not be deadlined, since there are no cracks in the windshield greater than 2 inches in length on the driver's side of the vehicle. Technician B says that any vision distorting defect in the windshield glass must be within the sweep of the wiper on the driver's side to deadline the vehicle. Who is correct?

TASK B.15

A. A only

B. B only

C. Both A and B

D. Neither A nor B

Answer A is incorrect. Commercial Vehicle Safety Alliance (CVSA) out-of-service criteria indicate cracks of ¼ inch in length or more.

Answer B is correct. Only Technician B is correct. Check the windshield glass for cracks, dirt, illegal stickers, or discoloration that obstructs the driver's view. Any crack over ¼-inch wide, intersecting cracks, discoloration not applied in manufacture, or other vision-distorting matter in the sweep of the wiper on the driver's side of the windshield will take a vehicle out of service by the commercial CVSA out-of-service standards.

Answer C is incorrect. Only Technician B is correct.

Answer D is incorrect. Technician B is correct.

11. A technician notices a whitish milky substance when changing the fluid in an axle. What is the most likely cause of this evidence?

TASK D.2.19

A. Common condensation

B. Infrequent driving and short trips

C. The axle being submerged in water

D. The vehicle being frequently driven during rainy or wet conditions

Answer A is incorrect. It is normal for drive axles to acquire slight condensation, which usually "boils off" during normal driving conditions.

Answer B is correct. Infrequent driving and short trips do not bring the axle to operating temperature and moisture does not burn off.

Answer C is incorrect. Although submerging the axles in water can introduce water into the axle, it is not the most likely cause.

Answer D is incorrect. Any water that happens to make its way into the axle would normally be "boiled off" during normal driving conditions.

TASK B.6

12. Technician A says that VORAD stands for variable output relay accessory device. Technician B says that anti-collision devices are now mandatory on all heavy-duty vehicles. Who is correct?

   A.   A only

   B.   B only

   C.   Both A and B

   D.   Neither A nor B

Answer A is incorrect. VORAD stands for vehicle on-board radar.

Answer B is incorrect. Anti-collision devices are not mandatory. Because anti-collision protection is high on the priority list of the Federal Highway Administration, collision warning systems will most likely become mandatory on heavy-duty vehicles in the near future.

Answer C is incorrect. Neither Technician is correct.

Answer D is correct. Neither Technician is correct. VORAD (vehicle on-board radar) is a collision warning system (CWS) that uses radar technology to sense some specific vehicle danger conditions. VORAD is used in trucks to sense straight-ahead closing velocities and proximity as well as blindside proximity. The system is electronically managed and consists of relatively few components. It must be connected into the vehicle electrical and electronic system so it can share sensor signals such as the input from the vehicle road speed sensor.

TASK C.11

13. When checking headlight aiming, Technician A says that the vehicle should be on level ground at a distance of 35 feet from a screen. Technician B says you must always use headlight aiming equipment when adjusting headlights. Who is correct?

   A.   A only

   B.   B only

   C.   Both A and B

   D.   Neither A nor B

Answer A is incorrect. The vehicle should be on level ground at a distance of 25 feet, not 35 feet as stated.

Answer B is incorrect. You do not always use headlight aiming equipment when adjusting headlights. One can use a screen or chart.

Answer C is incorrect. Neither Technician is correct.

Answer D is correct. Neither Technician is correct. When using aiming equipment, follow the instructions provided by the equipment manufacturer. Where headlight aiming equipment is not available, headlight aiming can be checked by projecting the upper beam of each light upon a screen or chart at a distance of 25 feet ahead of the headlights. The vehicle should be exactly perpendicular to the chart.

TASK C.8

14. The result of a shorted field winding inside an alternator rotor is:

   A.   No magnetic strength around the rotor.

   B.   Low current flow through the field winding.

   C.   Excessive alternator voltage output.

   D.   Damage to the voltage regulator.

Answer A is incorrect. There is still magnetic strength around the rotor.

Answer B is incorrect. The shorted condition causes high current flow through the field winding.

Answer C is incorrect. A shorted field winding in the rotor does not cause excessive alternator voltage output.

Answer D is correct. The shorted condition causes high current flow through the field winding and the voltage regulator, resulting in regulator damage.

15. Technician A checks coolant hoses for cracks and hardness every time a general inspection is made. Technician B says that a general coolant system inspection should be made every time a vehicle is scheduled for a PMI. Who is correct?

TASK A.21

   A. A only

   B. B only

   C. Both A and B

   D. Neither A nor B

Answer A is incorrect. Technician B is also correct.

Answer B is incorrect. Technician A is also correct.

Answer C is correct. Both Technicians are correct. Coolant hoses should be checked for cracks, hardness, deterioration, chafing, and bulges every time a general inspection is performed. All hose clamps should be checked for tightness and/or deterioration. Check for leaks around hoses and clamps both when cold and at operating temperature. A general coolant system inspection should be made every time a vehicle is scheduled for a PMI. Inspect the radiator for airflow restriction, leaks, damage, and mountings. Check operation of the fan clutch and inspect the fan assembly and shroud. Pressure test the cooling system and radiator cap. Inspect the coolant recovery system and water pump. Check the coolant for contamination, supplemental coolant additives (SCAs), and protection level. Service coolant filter/conditioner and drain/refill cooling system if scheduled.

Answer D is incorrect. Both Technicians are correct.

16. When performing a hydrometer test on a truck battery, all of the following are incorrect EXCEPT:

TASK C.3

   A. If the battery temperature is 50°F (10°C), the battery should be brought to room temperature before testing.

   B. If the battery temperature is 120°F (49°C), 0.020 specific gravity points should be subtracted from the hydrometer reading.

   C. The maximum variation in cell hydrometer readings is 0.100 specific gravity points.

   D. The battery is fully charged if all the cell hydrometer readings are above 1.265.

Answer A is incorrect. The battery does not need to be brought to room temperature before testing.

Answer B is incorrect. If the battery temperature is 120°F (49°C), 0.016 specific gravity point should be added to the hydrometer reading.

Answer C is incorrect. The maximum variation in cell hydrometer readings is 0.050 specific gravity point.

Answer D is correct. The battery is fully charged if all the cell hydrometer readings are above 1.265.

17. When discussing a battery capacity test with the battery at 70°F (21°C), Technician A says the battery discharge rate is two times the battery reserve capacity rating. Technician B says the battery is satisfactory if the voltage remains above 9.2 V. Who is correct?

    A.　A only

    B.　B only

    C.　Both A and B

    D.　Neither A nor B

    Answer A is incorrect. The battery discharge rate is one-half the battery CCA rating.

    Answer B is incorrect. The battery is satisfactory if the voltage remains above 9.6 V

    Answer C is incorrect. Neither Technician is correct.

    Answer D is correct. Neither Technician is correct. Determine the amperage rating of the vehicle (check manufacturer's specifications) and confirm that the proper batteries are installed in the vehicle. If the voltage is below 12.4 volts DC at 70°F, the battery should be recharged to above 12.6 volts. Apply a load through the tester at one-half the cold cranking amperes (CCA) for 15 seconds. If the voltage drops below 9.6 volts at 70°F, the battery is unserviceable.

18. A viscous vibration damper is being inspected. What would be the most serious consequence of a visible dent in the housing?

    A.　Engine crankshaft failure

    B.　Engine misfire

    C.　Engine surging

    D.　Knocking noise

    Answer A is correct. A damaged or failed vibration damper can cause crankshaft failure. The primary vibration damper's function is to reduce crankshaft torsional vibration. The damper housing is coupled to the crankshaft and uses springs, rubber, or viscous fluid to drive the inertia ring. Visually inspect the damper housing, noting any dents or signs of warpage. Check for indications of fluid leakage with damper in place. Trace evidence of leakage justifies replacement of the damper. With the damper on the crankshaft, check for wobble using a dial indicator. Viscous dampers should be checked for nicks, cracks, or bulges as well.

    Answer B is incorrect. A visible dent in the vibration damper housing will not cause engine misfire.

    Answer C is incorrect. The most serious consequence would be engine crankshaft failure.

    Answer D is incorrect. The most serious consequence would be engine crankshaft failure.

19. Technician A says to inflate mounted tires in a safety cage or using a portable lock ring guard. Technicians B says to first mount the tire on the truck and then inflate to the proper tire pressure. Who is correct?

    A.　A only

    B.　B only

    C.　Both A and B

    D.　Neither A nor B

    Answer A is correct. Only Technician A is correct. You must inflate mounted tires in a safety cage or using a portable lock ring guard.

    Answer B is incorrect. You do not mount the tire on the truck before inflation because an improperly fitted rim ring could dislodge, causing injury.

    Answer C is incorrect. Only Technician A is correct.

    Answer D is incorrect. Technician A is correct.

20. After removing a 30-ampere circuit breaker from a fuse panel, a technician tests breaker continuity with an ohmmeter. Technician A says the current from the ohmmeter will open the circuit breaker. Technician B says the ohmmeter should read infinity if the circuit breaker is satisfactory. Who is correct?

TASK C.10

A. A only

B. B only

C. Both A and B

D. Neither A nor B

Answer A is incorrect. The current from the ohmmeter will not open the circuit breaker.

Answer B is incorrect. The ohmmeter should provide a very low reading if the circuit breaker is satisfactory.

Answer C is incorrect. Neither Technician is correct.

Answer D is correct. Neither Technician is correct. Follow proper testing procedures for circuit protection devices outlined by either the vehicle or component manufacturer.

2013 © Delmar, Cengage Learning

21. An audible air leak is observed that appears to be coming from the front service brake chamber in the illustration above during an annual inspection. Which of the following is the LEAST LIKELY cause?

TASK D.1.14

A. Loose mounting bolts

B. Ruptured diaphragm

C. Loose chamber clamp

D. Missing clamp bolt

Answer A is correct. Loose mounting bolts would cause the chamber to move when the brakes are applied, but would be unlikely to cause an audible leak.

Answer B is incorrect. A ruptured diaphragm could cause an air leak.

Answer C is incorrect. A loose chamber clamp could cause an air leak.

Answer D is incorrect. A missing clamp bold could cause an air leak.

22. The current flow in an electric circuit is lower than specified. Technician A says the voltage supplied to the circuit may be higher than specified. Technician B says the resistance in the circuit may be higher than specified. Who is correct?

TASK C.1

A. A only

B. B only

C. Both A and B

D. Neither A nor B

Answer A is incorrect. To cause low current flow, the voltage supplied to the circuit may be lower than specified.

Answer B is correct. Only Technician B is correct. Higher than specified resistance cause low current flow.

Answer C is incorrect. Only Technician B is correct.

Answer D is incorrect. Technician B is correct.

TASKS D.4.12,
D.4.13

23. While discussing van and over-top trailer bodies, Technician A says if one floor cross member is broken and sagging below the lower rail, the vehicle should be tagged out of service. Technician B says if a drop frame is visibly twisted, the vehicle should be placed out of service. Who is correct?

   A. A only

   B. B only

   C. Both A and B

   D. Neither A nor B

   Answer A is incorrect. Three floor cross members have to be broken and sagging below the lower rail for the vehicle to be tagged out of service.

   Answer B is correct. Only Technician B is correct. If a drop frame is visibly twisted, the vehicle should be placed out of service.

   Answer C is incorrect. Only Technician B is correct.

   Answer D is incorrect. Technician B is correct.

TASK D.2.7

24. While inspecting a transmission for leaks, the technician notices that the gaskets appear to be blown out of their mating sealing surfaces. What should the technician check first?

   A. The transmission breather

   B. The shifter cover

   C. The release bearing

   D. The rear seal

   Answer A is correct. When the breather is plugged, internal pressure can build inside the transmission and push the oil out.

   Answer B is incorrect. The shift cover is an exit route for the pressurized oil, not the source of the pressure.

   Answer C is incorrect. The release bearing is not an exit route for pressurized oil, nor is it the source of the pressure.

   Answer D is incorrect. The rear seal is an exit route for the pressurized oil, not the source of the pressure.

TASK B.7

25. A digital electronic service tool (EST) is connected to a truck databus during a routine PMI to read engine data. Technician A says active trouble codes cannot be erased. Technician B says that an event audit trail can be read. Who is correct?

   A. A only

   B. B only

   C. Both A and B

   D. Neither A nor B

   Answer A is incorrect. Technician B is also correct.

   Answer B is incorrect. Technician A is also correct.

   Answer C is correct. Both Technicians are correct. Active trouble codes cannot be erased. The event audit trail codes can be read.

   Answer D is incorrect. Both Technicians are correct.

26. Technician A says any type of wire may be used to replace a burned fusible link as long as the gauge is one size smaller than the wire in the circuit being protected. Technician B says that a circuit breaker should be replaced after repairing a short in a circuit. Who is correct?

TASK C.7

    A. A only

    B. B only

    C. Both A and B

    D. Neither A nor B

Answer A is incorrect. A burned fusible link must be replaced with a length of the specified fusible link wire.

Answer B is incorrect. A circuit breaker resets itself and does not have to be replaced after repairing a short in a circuit.

Answer C is incorrect. Neither Technician is correct.

Answer D is correct. Neither Technician is correct. Follow recommended procedures when replacing any component or wiring. Never alter a circuit or use improper repair techniques. A circuit breaker is designed not to fail when subjected to a circuit overload. When a circuit draws excessive amperage, the circuit breaker is designed to open the circuit.

2013 © Delmar, Cengage Learning

27. What is the item indicated by the arrow in the illustration above?

TASK A.4

    A. A water pump pulley

    B. An air pump pulley

    C. A fuel pump pulley

    D. An idler pulley

Answer A is incorrect. A water pump pulley would be above the vibration damper, not to the side of it.

Answer B is incorrect. An air pump is not part of a diesel engine assembly.

Answer C is incorrect. The fuel pump is not belt-driven.

Answer D is correct. An idler pulley is shown in the illustration.

TASK D.1.15

28. Technician A says that air drum brake linings should be replaced when there are less than ¼ inch of linings when measured at the center of the shoe. Technician B says that brake linings should be replaced when they are saturated with oil or grease. Who is correct?

    A. A only

    B. B only

    C. Both A and B

    D. Neither A nor B

    Answer A is incorrect. Technician B is also correct.

    Answer B is incorrect. Technician A is also correct.

    Answer C is correct. Both Technicians are correct. Linings with a thickness less than ¼ inch at the shoe center for air drum brakes should be replaced. Brake linings that are saturated with oil, grease, or brake fluid should be replaced. Linings and pads should be firmly attached to the shoe. Unequal lining wear should be noted and the cause investigated. When replacing linings, both wheels of an axle should be replaced to maintain brake balance. The frictional rating of linings is coded by letter codes stamped on the edge of the lining. Combination sets of shoe linings are occasionally used, and these linings use different friction ratings on the primary and secondary shoes. When combination friction lining sets are used, care should be taken to install the lining blocks in the correct locations on the brake shoes.

    Answer D is incorrect. Both Technicians are correct.

TASK A.18

29. Which of the following cause most heavy-duty pneumatic fan clutch failures?

    A. Seized support bearings

    B. Air leaks

    C. Electrical shorts

    D. Obstructions

    Answer A is incorrect. The hub bearings generally do not fail prematurely.

    Answer B is correct. Most failures are caused by air leaks.

    Answer C is incorrect. An electrical problem will cause the clutch to shut down, a failsafe device.

    Answer D is incorrect. Obstructions generally cause immediate damage to the fan blades.

30. Technician A says to block the wheels before adjusting the service brakes on a truck. Technician B says that the air reservoir pressure must be dropped below the cut-in value before the air compressor will resume its effective cycle. Who is correct?

    A. A only

    B. B only

    C. Both A and B

    D. Neither A nor B

TASKS D.1.2,
D.1.17

Answer A is incorrect. Technician B is also correct.

Answer B is incorrect. Technician A is also correct.

Answer C is correct. Both Technicians are correct. You do block the wheels before attempting to adjust service brakes. The air reservoir pressure must be dropped below cut-in pressure before effective cycle resumes. Air brake system pressure in most trucks is set at a value between 110 and 130 psi (758 and 896 kPa), with 120 psi (827 kPa) being typical. System pressure is known as cut-out pressure, the pressure at which the governor outputs the unloader signal to the compressor. The unloader signal is maintained until pressure in the supply tank drops to the cut-in value. Cut-in pressure is required by FMVSS 121 to be no more than 25 psi (172 kPa) less than the cut-out value. The difference on most systems ranges between 20 and 25 psi (137 and 172 kPa). To check air governor cut-in pressure, with the engine running, build system pressure to achieve cut-out. Next, pump the service brake foot valve to drop pressure. Listen for the moment the compressor resumes loaded cycle; this is governor cut-in.

Answer D is incorrect. Both Technicians are correct.

31. During a PMI, the technician tests the air supply system and finds that buildup time is slow. Which of the following is the most likely cause?

    A. A defective air governor

    B. A leak in a service brake chamber

    C. An air leak in the cab

    D. A defective air compressor

TASK D.1.10

Answer A is incorrect. A defective air governor will not cause a slow build-up, only cut-in and cut-off problems. The governor manages system pressure. It monitors pressure in the supply tank by means of a line directly to it. The governor manages compressor loaded and unloaded cycles. Loaded cycle is the compressor effective cycle; that is, the compressor is pumping air. The unloaded cycle occurs when the compressor is driven by the engine but it is not actually compressing air. The compressor is in the loaded cycle until it receives an air signal from the governor to put it into the unloaded cycle. The governor simply controls whether the compressor is in the loaded or unloaded operating mode. Air brake system pressure (cut-out pressure) in most trucks is set at values between 110 and 130 psi (758 and 896 kPa) with 120 psi (827 kPa) being typical. Cut-in pressure is required by FMVSS 121 to be no more than 25 psi (172 kPa) less than the cut-out value.

Answer B is incorrect. A leak in the service air brake chamber will not cause slow build-up because it is on the application side, not the supply side.

Answer C is incorrect. An air leak in the cab will not cause slow build-up since the air supply is protected by pressure-protection valves. A pressure-protection valve traps air in the air brake system if a leak occurs in one of the air-operated devices on the truck. When the pressure in air-operated devices is equal to the reservoir pressure, the pressure-protection valve remains open. If an air leak occurs in one of the air-operated devices, the pressure in the device drops below reservoir pressure. Under this condition, the pressure-protection valve closes to prevent any further escape of air.

Answer D is correct. A defective air compressor is the most likely cause of slow build-up.

TASK B.20

32. The trailer lights of a double combination vehicle are being inspected. Technician A says that the wiring connector for all trailer lights is usually an 8-pin connector. Technician B says that it is mandatory to check the lights on the last trailer first in a multiple combination vehicle. Who is correct?

A.　A only

B.　B only

C.　Both A and B

D.　Neither A nor B

Answer A is incorrect. The most common method to link the tractor's light system to the trailer's and power various lights on a trailer requires the use of a seven-way receptacle and trailer cord plug mated together.

Answer B is incorrect. A circle-check or pre-trip inspection begins on the left front side of the truck and continues in a counterclockwise direction around it.

Answer C is incorrect. Neither Technician is correct.

Answer D is correct. Neither Technician is correct. Many vehicle manufactures use colored wires to more easily identify common electrical circuits. An adaptation of the CVSA would use the following step process: vehicle overview, engine compartment, inside the cab (drivers side), front of cab, left side of cab, left fuel tank area, left rear tractor area, left side of trailer, rear of trailer, right side of trailer, right rear tractor area, right fuel tank area, right side of cab, cab.

TASK A.9

33. A vehicle is being inspected for a safety certification. Technician A states if wetness is seen around a seam of a fuel tank without forming any drops, it is passable. Technician B says that it is not necessary to drain water from the fuel when performing an inspection during the summer months. Who is correct?

A.　A only

B.　B only

C.　Both A and B

D.　Neither A nor B

Answer A is incorrect. No leakage or wetness of fuel is permitted on or around fuel tanks, mountings, fuel lines, or caps. When fuel is drawn out of a tank, it is replaced by air. This air always contains some percentage of moisture and in areas of high humidity, this can be extreme. Cooling temperatures cause this vaporized moisture to condense as water, which settles at the bottom of the tank. It is good practice to keep fuel tanks full, especially after the vehicle has been run and the tanks and fuel are warm. This keeps moisture-laden air out of the tanks and reduces water-in-fuel problems.

Answer B is incorrect. The time of season has no bearing on draining water from fuel tanks.

Answer C is incorrect. Neither Technician is correct.

Answer D is correct. Neither Technician is correct. The fuel tank mounts should be securely attached to the frame and the tank free of any cracks, corrosion, and damage that could cause future failure. Check to ensure that the fuel tank cap seal is in place and in good condition. The fuel lines, including the fuel crossover line must be leak free and should be secured by loom or brackets. Fuel lines should be checked for chafing and the possibility of future leaks or breakdowns.

2013 © Delmar, Cengage Learning

34. Which of these conditions do the shoes in the illustration above show?

A. Unequal wear

B. Poor contact at center of linings

C. Lining material tapered

D. Normal wear

TASK D.1.15

Answer A is correct. The illustration shows unequal shoe lining wear.

Answer B is incorrect. Poor contact at center would have lining ends worn excessively.

Answer C is incorrect. This would appear to gradually decrease in thickness from one end to the other.

Answer D is incorrect. This would indicate a fairly uniform wear appearance.

35. Technician A says overfilling a manual transmission could result in overheating of the transmission. Technician B says overfilling a transmission could cause excessive aeration of the transmission fluid. Who is correct?

A. A only

B. B only

C. Both A and B

D. Neither A nor B

TASK D.2.17

Answer A is incorrect. Technician B is also correct.

Answer B is incorrect. Technician A is also correct.

Answer C is correct. Both Technicians are correct. Overfilling a manual transmission could result in overheating of the transmission. Overfilling a manual transmission could cause excessive aeration of the transmission fluid.

Answer D is incorrect. Both Technicians are correct.

36. When checking for diesel engine fuel system leaks, the engine must be:

A. Cold.

B. At normal operating temperature.

C. Running.

D. Off.

TASK A.5

Answer A is incorrect. The temperature of the engine is not important for this test.

Answer B is incorrect. The temperature of the engine is not important for this test.

Answer C is correct. When checking for fuel leaks on an engine, it is mandatory that the engine be operating so the fuel system is pressurized, thus making a leak very apparent.

Answer D is incorrect. The engine must be running to pressurize the fuel system for the test.

TASK E.5

37. Technician A says that an inspection of the road speed governor is required during the pre-trip inspection. Technician B says that more emphasis should be placed on the post-trip inspection rather than the pre-trip inspection. Who is correct?

A. A only

B. B only

C. Both A and B

D. Neither A nor B

Answer A is incorrect. The operation of a road speed governor cannot be checked on a pre-trip inspection.

Answer B is incorrect. Although the regulations seem to place greater emphasis on the post- trip inspection, most operators agree the pre-trip inspection is most important.

Answer C is incorrect. Neither Technician is correct.

Answer D is correct. Neither Technician is correct. Checking the road speed limiter is part of a vehicle road test. Connect the appropriate engine scan tool, original equipment manufacturer (OEM) special tool, or laptop computer to the truck's diagnostic connector. During the road test, activate the tool's parameter specification readout and record the road speed. Reprogram if required. Because the driver doing the inspection will be operating that particular vehicle, the incentive to ensure that the vehicle is safe is probably greater. Repairs are cheaper and less time-consuming if the driver finds them on the pretrip before inspectors levy fines or impose penalties.

TASK D.2.11

38. Fluid leaks are evident between the axle housing and the carrier assembly. What would the most likely cause of this leak be?

A. Damaged gasket or missing sealant

B. Repeated overloading of the drive train

C. Plugged axle housing breather vent

D. Moisture contaminated axle lubricant

Answer A is correct. A damaged gasket or missing sealant could cause a leak in this location.

Answer B is incorrect. Overloading of the drive train would increase the axle temperature and thin out the fluid, but a good gasket should still seal this area.

Answer C is incorrect. A plugged breather would increase the pressure inside the axle assembly, but this condition usually forces oil past the lips of the wheel seal, which is the weakest sealing point.

Answer D is incorrect. Moisture contamination will have no effect on the ability of gaskets to seal or leak.

2013 © Delmar, Cengage Learning

39. Technician A says that the dash control button on the left in the illustration above is the trailer service brake valve. Technician B says that the dash control button on the right in the same illustration is the system park valve. Who is correct?

TASK D.1.1

   A.   A only

   B.   B only

   C.   Both A and B

   D.   Neither A nor B

Answer A is incorrect. The valve on the left is the trailer air supply valve. It is octagonal and red in color. The trailer air supply valve is used to supply air to apply and release the trailer parking brakes. The valve is pulled outward to vent air pressure in the trailer parking chambers and to allow the chamber springs to apply the parking brakes. When the trailer air supply valve is pushed inward, air pressure is supplied through the trailer supply line to the trailer parking brake chambers to release the parking brakes.

Answer B is correct. Only Technician B is correct. The valve on the right in the illustration is the system park valve. It is diamond shaped and yellow in color. The park control (system park) valve is used to apply and release the tractor and trailer parking brakes. If the park control valve is pushed inward, the parking brakes are released, and pulling the park control valve outward results in parking brake application.

Answer C is incorrect. Only Technician B is correct.

Answer D is incorrect. Technician B is correct.

40. What circuit is being operated when a starter draw test is being performed?

TASKS
C.5, C.6

   A.   Charging circuit

   B.   Cranking circuit

   C.   Regulator circuit

   D.   Control circuit

Answer A is incorrect. The charging circuit is not related to the starting circuit.

Answer B is correct. The cranking circuit is involved in a starter draw test. The starter draw test is used to measure the amount of current in amperes that the starter circuit draws to crank the engine. This information is used to determine the next step needed to pinpoint a starter system problem.

Answer C is incorrect. The regulator circuit is not related to the starter circuit.

Answer D is incorrect. The control circuit is not related to the starter circuit.

Steering knuckle

Circular clip

Hub

Constant velocity universal joint

Wheel bearing assembly

Washer

Hub nut

2013 © Delmar, Cengage Learning

TASKS D.3.8, D.3.7

41. The bearings in the wheel hub shown in the illustration above:

A. Must be lubricated with wheel bearing grease before assembly.

B. Are permanently lubricated with the specified oil.

C. Require a specific hub end-play.

D. Require inner and outer hub seals.

Answer A is incorrect. The bearings in the illustration are lubricated with the specified oil.

Answer B is incorrect. The bearings are not permanently lubricated with the specified oil.

Answer C is correct. The bearings require a specific hub end-play.

Answer D is incorrect. The bearings only have an inner hub seal.

TASK E.2

42. During a PMI, the antilock brake system (ABS) warning light is illuminated with the engine running. Technician A says there is a mechanical defect in the ABS. Technician B says the ABS will still function normally. Who is correct?

A. A only

B. B only

C. Both A and B

D. Neither A nor B

Answer A is incorrect. There is an electrical defect in the ABS.

Answer B is incorrect. The ABS will be inoperative and normal non-ABS braking is maintained.

Answer C is incorrect. Neither Technician is correct.

Answer D is correct. Neither Technician is correct. During the road/operational test, check for proper operation of all instrumentation, gauges, and indicator lights, document any inoperative components, troubleshoot the cause, and perform proper repairs.

2013 © Delmar, Cengage Learning

43. Technician A says the device in the illustration above is to restrain the tire while it is being inflated. Technician B says that the device could be used to center the tire for reuse. Who is correct?

TASK D.3.14

    A. A only

    B. B only

    C. Both A and B

    D. Neither A nor B

Answer A is correct. Only Technician A is correct. The device in the illustration is to restrain the tire while it is being inflated.

Answer B is incorrect. Once a tire's shape has been altered (belt has shifted, high speed distortion, etc.), there is no way to reshape the rubber structure back to its original form.

Answer C is incorrect. Only Technician A is correct.

Answer D is incorrect. Technician A is correct.

44. What can result from a shock absorber outer tube touching the fluid reservoir?

TASK D.4.2

    A. Premature bushing wear

    B. Scraping noise with travel

    C. Main shaft seal failure

    D. Excessive fatigue in mount assembly

Answer A is incorrect. A shock absorber outer tube touching the fluid reservoir will not cause premature bushing wear.

Answer B is correct. The shock absorber outer tube touching the fluid reservoir will cause a scraping noise during shock travel.

Answer C is incorrect. A shock absorber outer tube touching the fluid reservoir will not cause main shaft seal failure.

Answer D is incorrect. A shock absorber outer tube touching the fluid reservoir will not cause excessive fatigue in the mount assembly.

TASKS A.23, A.25

45. Technician A says that when refilling a cooling system it may be necessary to open a bleed valve. Technician B says that when checking a cooling system that uses ELC, to check the coolant conditioner level by using an SCA test strip specifically designed for ELC coolants. Who is correct?

    A.   A only
    B.   B only
    C.   Both A and B
    D.   Neither A nor B

    Answer A is correct. Only Technician A is correct. It is necessary to ensure that air is bled from a cooling system when refilling the system. This can be done by opening a hose or fitting on the top of the engine before the thermostat and filling the system until liquid coolant appears. This prevents trapped air from creating hot spots when the engine is first started. Some systems employ bleed valves to aid in removing trapped air from the cooling system.

    Answer B is incorrect. ELC coolants do not require in-service SCA conditioner monitoring. No inhibitor test kits are required to monitor the ELC SCA level. If a coolant filter is used, replace the existing filter with an SCA-free filter. Extended-life coolants promise a service life of 600,000 miles or six years with one additive recharge at 300,000 miles or three years. This compares with a typical service life of two years during which up to 20 recharges of SCA would be required for conventional EG and PGs.

    Answer C is incorrect. Only Technician A is correct.

    Answer D is incorrect. Technician A is correct.

TASK D.4.14

46. A scheduled inspection is being performed. Technician A says that if lift gates and controls are not on the checklist, then it is not mandatory to inspect them. Technician B says you always check cargo ramps on a vehicle equipped with them. Who is correct?

    A.   A only
    B.   B only
    C.   Both A and B
    D.   Neither A nor B

    Answer A is incorrect. Generally, if any components or systems are suspected of requiring servicing, those non-checklisted areas are more thoroughly inspected and/or serviced immediately.

    Answer B is correct. Only Technician B is correct. You always check the cargo ramps on a vehicle equipped with them during a PMI.

    Answer C is incorrect. Only Technician B is correct.

    Answer D is incorrect. Technician B is correct.

TASK D.4.10

47. The vehicle to which a pintle hook or drawbar is attached should be tagged out of service for any of the these defects EXCEPT:

    A.   Wear in the pintle hook horn that exceeds 10 percent of the horn section.
    B.   Loose or missing mounting bolts.
    C.   Any welded repairs to the drawbar eye.
    D.   Cracks in the pintle hook mounting surface.

    Answer A is correct. The vehicle is not tagged out of service unless wear in the pintle hook horn exceeds 20 percent of the horn section.

    Answer B is incorrect. The vehicle is tagged out of service if there are loose or missing pintle hook mounting bolts.

    Answer C is incorrect. The vehicle is tagged out of service if there are any welded repairs to the drawbar eye.

    Answer D is incorrect. The vehicle is tagged out of service if there are cracks in the pintle hook mounting surface.

48. A truck's clutch is being measured for its free travel. Technician A says the measurement should be between 1 and 2 inches. Technician B says that on a hydraulic clutch, there should be no free travel. Who is correct?

    A. A only

    B. B only

    C. Both A and B

    D. Neither A nor B

    TASK E.1

    Answer A is correct. Only Technician A is correct. The measurement should be between 1 and 2 inches.

    Answer B is incorrect. When free travel disappears, clutch failure can occur rapidly.

    Answer C is incorrect. Only Technician A is correct.

    Answer D is incorrect. Technician A is correct.

49. What would be the LEAST LIKELY cause of a front wheel shimmy?

    A. Kingpin play

    B. Improperly adjusted steering gear relief plungers

    C. Loose Pitman arm

    D. Tire imbalance

    TASKS D.3.1, D.3.5, D.3.6, D.3.7

    Answer A is incorrect. Excessive kingpin causes front wheel shimmy.

    Answer B is correct. Steering gear relief plungers are the least likely cause of a front wheel shimmy because they are only a factor at steering lock.

    Answer C is incorrect. Looseness in steering linkages can cause wheel shimmy.

    Answer D is incorrect. Wheel shimmy can be caused by tire imbalance.

50. Technician A says that a proper inspection of the cooling system should include a check of coolant-usage records. Technician B changes the coolant at every PMI. Who is correct?

    A. A only

    B. B only

    C. Both A and B

    D. Neither A nor B

    TASKS A.5, A.25

    Answer A is correct. Only Technician A is correct. While checking the cooling system, checking the records of coolant usage is important to indicate whether excess coolant is being lost. A proper inspection of the cooling system should include a check of the cooling system maintenance history, coolant change intervals, component replacement(s), and past diagnoses.

    Answer B is incorrect. The coolant is changed too frequently and does not take into consideration factors such as whether extended-life coolant is being used, the time interval between the last cooling system drain and refill, or what scheduled PMI is being performed. Identify the coolant being used and be sure to follow the pertinent change intervals and refill with the correct type. Extended life coolants (ELCs) have a longer service life than ethylene glycol (EG) and propylene glycol (PG) coolants.

    Answer C is incorrect. Only Technician A is correct.

    Answer D is incorrect. Technician A is correct.

PREPARATION EXAM 4—ANSWER KEY

1.	A	**18.**	A	**35.**	B
2.	B	**19.**	D	**36.**	C
3.	C	**20.**	A	**37.**	C
4.	A	**21.**	D	**38.**	C
5.	A	**22.**	A	**39.**	C
6.	A	**23.**	D	**40.**	A
7.	D	**24.**	C	**41.**	C
8.	D	**25.**	D	**42.**	B
9.	B	**26.**	C	**43.**	C
10.	C	**27.**	A	**44.**	C
11.	A	**28.**	C	**45.**	A
12.	A	**29.**	D	**46.**	A
13.	C	**30.**	C	**47.**	A
14.	C	**31.**	B	**48.**	D
15.	A	**32.**	A	**49.**	D
16.	B	**33.**	D	**50.**	D
17.	C	**34.**	A		

PREPARATION EXAM 4—EXPLANATIONS

TASKS D.2.3, D.2.4, B.25

1. A truck mechanically actuated clutch is being measured for freeplay. Technician A says the measurement should be between 1 and 2 inches. Technician B says that on a hydraulic clutch freeplay should be 2 to 3 inches. Who is correct?

   A. A only

   B. B only

   C. Both A and B

   D. Neither A nor B

   Answer A is correct. Only Technician A is correct. The measurement should be between 1 and 2 inches with 1½ inches being typical. With the clutch pedal fully raised, there should always be ⅛ inch freeplay between the release yoke fingers and the clutch release bearing pads. This freeplay is taken up by the first 1 to 1½ inches of pedal travel. Measure the distance the pedal can be pushed down before thrust is exerted on the clutch release bearing to check clutch-pedal free travel.

   Answer B is incorrect. The manufacturer's recommendations should be checked on hydraulic clutches.

   Answer C is incorrect. Only Technician A is correct.

   Answer D is incorrect. Technician A is correct.

2. Which of the following defines the cloud point of diesel fuel?

    A. The lowest temperature at which it will flow

    B. The temperature at which a haze of wax crystals will form

    C. The highest temperature before evaporation begins

    D. The temperature at which the fuel produces black smoke

TASK A.9

Answer A is incorrect. The pour point of a fuel is generally considered to be 5 °F above the temperature at which the fuel will no longer flow. As the temperature is lowered below the cloud point, more and more of the paraffinic compounds of the fuel crystallize. As they do, they begin to combine to form a solid gel-like structure, which cannot be poured or pumped. Pour point depressant additives are required for extreme cold weather operation in #1D and #2D diesel fuel. Pour point depressants have no effect on cloud point.

Answer B is correct. The cloud point is the temperature at which the normal paraffin in a fuel becomes less soluble and begins to precipitate as wax crystals and make the fuel appear cloudy and hazy. The cloud point exceeds the pour point by 5 °F (3 °C) to 25 °F (15 °C).

Answer C is incorrect. The point at which evaporation begins is not a factor to consider in normal and/or severe operating conditions.

Answer D is incorrect. Black smoke is produced when excess fuel is burned.

3. Technician A says you turn the key on the ignition to perform electronic system self-checks. Technician B says the ignition circuit powers most computer-controlled chassis systems. Who is correct?

    A. A only

    B. B only

    C. Both A and B

    D. Neither A nor B

TASKS B.1,
B.1.7

Answer A is incorrect. Technician B is also correct.

Answer B is incorrect. Technician A is also correct.

Answer C is correct. Both Technicians are correct. You turn the key on the ignition to perform electronic system self-checks. The ignition switch powers up most computer-controlled chassis systems.

Answer D is incorrect. Both Technicians are correct.

2013 © Delmar, Cengage Learning

TASK C.2

4. What type of connections are pictured in the illustration above?

   A.   Parallel

   B.   Parallel-series

   C.   Series

   D.   Series-parallel

Answer A is correct. Parallel connections are shown in the illustration.

Answer B is incorrect. Parallel-series connections would not be connected in this manner.

Answer C is incorrect. Series would have all the connections on both sides connected this way.

Answer D is incorrect. Series-parallel connections would have the ground side circuit in series.

TASK E.8

5. When performing a test drive on a vehicle, Technician A says to test the service brakes by applying and releasing them. Technician B says that although a test drive is not mandatory, it is a good idea to perform one after a vehicle has been serviced. Who is correct?

   A.   A only

   B.   B only

   C.   Both A and B

   D.   Neither A nor B

Answer A is correct. Only Technician A is correct. The service brakes are tested during road tests of the vehicle. The brakes that are activated by the brake pedal are referred to as the service brakes. Work the brakes during the road test using a mixture of snub braking and aggressive panic stops. The Federal Motor Vehicle Safety Standards (FMVSS) sets the standards used in the United States and Canada to define stopping distances as well as other braking safety criteria.

Answer B is incorrect. A test drive is a mandatory procedure of a PMI.

Answer C is incorrect. Only Technician A is correct.

Answer D is incorrect. Technician A is correct.

6. In addition to replacing the fuel filter and priming the fuel system, what else should be performed when doing a preventive maintenance inspection (PMI) on a fuel system?

    A. Check the fuel tank mountings.
    B. Test fuel pour point.
    C. Check heater lines for proximity to the fuel lines.
    D. Recalibrate the fuel gauge.

TASK A.9

Answer A is correct. The condition of fuel tanks, mountings, lines, and caps are a concern during a PMI. The fuel tank mounts should be securely attached to the frame and the tank and be free from any cracks, corrosion, and damage that could cause future failure. There should be no signs of tank shifting and all mounting bolts should be in place and tight. Tanks must be checked for any signs of leakage at any point, with special attention to all seams and any damaged areas. Check to ensure that the cap seal is in place and in good condition.

Answer B is incorrect. Pour point tests are never performed as part of a PMI.

Answer C is incorrect. The heater lines have nothing to do with a fuel system PMI.

Answer D is incorrect. Fuel gauges are not recalibrated as part of PMI.

7. An inspection is being performed on a heavy-duty truck brake system and grease soaked linings are found on three wheels of the trailer's rear tandem. Technician A says to inform the driver. Technician B writes up the problems and informs the driver. Who is correct?

    A. A only
    B. B only
    C. Both A and B
    D. Neither A nor B

TASK D.1.15

Answer A is incorrect. Informing the driver on such an important safety issue is not enough.

Answer B is incorrect. An important safety problem such as brakes requires the vehicle to be deadlined until proper repairs are made and the work is noted using a sign-off sheet before the vehicle is put back into service.

Answer C is incorrect. Neither Technician is correct.

Answer D is correct. Neither Technician is correct. CVSA out-of-service criteria shows that brake linings that are saturated with oil, grease, or brake fluid is an out-of-service condition.

8. When winterizing a truck, which of the following procedures should be done first?

    A. Replacing lost coolant
    B. Cleaning the coolant
    C. Adding antifreeze
    D. Testing the condition of the coolant

TASK A.23

Answer A is incorrect. Replacing the lost coolant with water or antifreeze would result in an incorrect reading and does not follow the proper sequence of servicing the cooling system.

Answer B is incorrect. Cleaning the coolant is performed after testing the coolant's condition.

Answer C is incorrect. Adding antifreeze to the system should only be performed after a full evaluation of the existing coolant has been performed.

Answer D is correct. Test the condition of the coolant first. Coolant freeze and boil protection levels can be tested with a hydrometer or a refractometer. Different antifreeze and water ratios are required to raise or lower the protection level depending upon the temperatures the vehicle will encounter. A 50/50 mixture is most common.

TASK D.3.18

9. Which of these criteria is the LEAST LIKELY to be considered tire mismatching?

   A. Radials and bias ply on the same axle
   B. Dual wheels of the same nominal size but different tread design
   C. Installation of a new tire next to a worn-out tire on the same axle
   D. Tires with different circumferences on the same axle

   Answer A is incorrect. Mounting tires of different construction on the same axle is mismatching. Radial-ply and bias-ply tires must not be mixed on the same axle. All the tires on an axle must be either radial ply or bias ply. On trucks with two or more drive axles, the tire matching on all the axles are necessary so all the tires are the same type. Because radial tires defect more than bias-ply tires under load, a mix of bias-ply and radial-ply tires overloads the bias-ply tires. This action causes premature tire tread wear and possible tire blowout.

   Answer B is correct. Though it is not desirable to mix tread designs, both tires are the same nominal dimension, so this is least likely.

   Answer C is incorrect. Installation of a new tire next to an old tire is considered mismatching.

   Answer D is incorrect. Mounting tires of different circumferences and diameters is always considered mismatching. Matching tire sizes on dual wheels prevents tire tread wear from slippage due to uneven surface areas. Dual drive tires may be measured with a tape measure, string gauge, square, straightedge, or caliper.

TASK A.3

10. Technician A says the vibration damper counterbalances the back-and-forth twisting motion of the crankshaft each time a cylinder fires. Technician B says if the seal contact area on the vibration damper hub is scored, the damper assembly must be replaced. Who is correct?

    A. A only
    B. B only
    C. Both A and B
    D. Neither A nor B

    Answer A is incorrect. Technician B is also correct.

    Answer B is incorrect. Technician A is also correct.

    Answer C is correct. Both Technicians are correct. The vibration damper counterbalances the back-and-forth twisting motion of the crankshaft each time a cylinder fires. If the seal contact area on the vibration damper hub is scored, the damper assembly must be replaced.

    Answer D is incorrect. Both Technicians are correct.

2013 © Delmar, Cengage Learning

Dial indicator

Micrometer

11. What is being performed in the illustration above?

TASK D.1.16

A. Lateral runout and rotor thickness

B. Wheel bearing runout

C. Rotor thickness and radial runout

D. Parallelism

Answer A is correct. Lateral runout and rotor thickness measurements are shown in the illustration. To check lateral runout, mount a dial indicator on the steering arm or anchor plate with the indicator plunger contacting the rotor 1 inch from the edge of the rotor. Check lateral runout on both sides of the rotor. Zero the indicator and measure the total indicated runout (TIR). The lateral runout TIR is always a tight dimension, seldom exceeding 0.015 inch. If lateral runout exceeds the specifications, the rotor will have to be either machined or replaced. Thickness variation should also be measured at 12 equidistant points with a micrometer at about 1 inch from the edge of the rotor. If thickness measurements vary by more than the specified maximum, usually around 0.002 inch, the rotor should be machined or replaced.

Answer B is incorrect. Lateral runout and rotor thickness measurements are shown in the illustration.

Answer C is incorrect. Lateral runout and rotor thickness measurements are shown in the illustration.

Answer D is incorrect. Lateral runout and rotor thickness measurements are shown in the illustration.

12. A starting motor fails to crank the engine and indicates a very high current draw. The most likely cause of this problem is:

TASK C.6

A. A direct ground in the starter field coils.

B. An open circuit in the armature windings.

C. Broken brush springs and no tension on the brushes.

D. Failed magnetic switch.

Answer A is correct. A direct ground in the starter field coils causes high current draw and no starter operation.

Answer B is incorrect. An open circuit in the armature windings causes no starter current draw.

Answer C is incorrect. Broken brush springs and no tension on the brushes causes low starter current draw.

Answer D is incorrect. Failed magnetic switch causes no starter current draw.

TASK B.9

13. A fleet manager is restructuring a PMI program. Technician A says that fire extinguisher distribution varies with the type of vehicle it is installed in. Technician B says that it is the fleet manager's responsibility to see that personnel performing annual inspections are qualified. Who is correct?

    A. A only

    B. B only

    C. Both A and B

    D. Neither A nor B

Answer A is incorrect. Technician B is also correct.

Answer B is incorrect. Technician A is also correct.

Answer C is correct. Both Technicians are correct. The type of fire extinguisher carried on a truck should be determined by factors such as the type of cargo carried. Most fire extinguishers are marked by a letter and symbol to indicate the classes of fires for which they can be used. Every truck or truck-tractor with a gross vehicle weight rating (GVWR) of 10,001 pounds or more must have a fire extinguisher. If the vehicle is used for transporting hazardous material that requires placards, the extinguisher must have an Underwriter's Laboratory (UL) rating of 10B:C or more. If the vehicle is not used for hazardous materials, an extinguisher with a UL rating of 5B:C may be used. The rating is usually shown near the UL certification on the extinguisher. Section 396.19 of the Federal Motor Carrier Safety Regulations states it is the fleet manager's responsibility to ensure that all personnel doing annual inspections are qualified to perform those inspections. They should understand the inspection criteria and be able to identify defective components and systems.

Answer D is incorrect. Both Technicians are correct.

TASK D.1.10

14. An air brake system has slower than specified air pressure buildup time. What could be the cause of this problem?

    A. A smaller than specified supply reservoir

    B. An air leak in the compressor air intake filter

    C. Carbon buildup in the compressor discharge hose

    D. An excessive amount of water in the supply reservoir

Answer A is incorrect. A smaller than specified supply reservoir causes a faster air pressure buildup time.

Answer B is incorrect. An air leak in the compressor air intake filter does not cause a longer-than-specified compressor air pressure buildup time.

Answer C is correct. Carbon buildup in the compressor discharge hose causes a longer-than-specified compressor air pressure buildup time.

Answer D is incorrect. An excessive amount of water in the supply reservoir causes a faster air pressure buildup time.

15. Visual inspection of the power steering belt(s) can reveal all the following EXCEPT:

TASK A.4

   A. Proper belt tension.
   B. Premature wear due to misalignment.
   C. Correct orientation of dual belt application.
   D. Proper belt seating in the pulley.

Answer A is correct. A visual inspection will not reveal the proper belt tension. A belt tension gauge should be used whenever possible for the most accurate readings. If one is not available, then the normal deflection points at the specified locations and measurements should be observed.

Answer B is incorrect. Premature wear due to misalignment can be seen on a visual inspection. Belt pulleys must be properly aligned to minimize belt wear. The edges of the pulleys must be in line when a straightedge is placed on the pulleys.

Answer C is incorrect. Correct orientation of a dual belt application can be seen on a visual inspection.

Answer D is incorrect. Proper belt seating in the pulley groove can be seen on a visual inspection. Because the friction surfaces are on the side of a V-belt, wear occurs in this area. If the belt edges are worn, the belt may be rubbing on the bottom of the pulley. The lower edge should never touch the bottom of the pulley groove.

16. A driver complains of a door not closing properly and has brought it in for service. Technician A immediately greases the hinges. Technician B retrieves the vehicle records and checks for past problems with the doors and makes a note of the driver reports. Who is correct?

TASKS B.21, B.23

   A. A only
   B. B only
   C. Both A and B
   D. Neither A nor B

Answer A is incorrect. Technician A immediately greases the hinges although that may not be the cause of the door not closing properly. Many factors could contribute to the door not closing properly.

Answer B is correct. Only Technician B is correct. Listening to or reading the driver's reports and then checking the truck's maintenance records for past problems or repairs before proceeding is the proper diagnostic sequence to follow. A thorough inspection of the door and associated components such as hinges and latches should be performed before determining a proper course of action.

Answer C is incorrect. Only Technician B is correct.

Answer D is incorrect. Technician B is correct.

TASK C.3

17. Technician A always begins battery inspection with a general visual first, followed by an open circuit voltage test. Technician B says that it may be necessary to remove the surface charge from a battery before it is open circuit voltage tested. Who is correct?

    A. A only

    B. B only

    C. Both A and B

    D. Neither A nor B

Answer A is incorrect. Technician B is also correct.

Answer B is incorrect. Technician A is also correct.

Answer C is correct. Both Technicians are correct. Begin battery service with a visual inspection looking for: loose hold-downs, defective cables, damaged terminal posts, loose terminal connections, clogged vents, corrosion, dirt or moisture, cracked case, and battery box condition. Following a visual inspection, a battery open circuit voltage test can be used to determine its state of charge. A battery that has recently been charged by running the engine or using a battery charger must have its surface charge removed before performing the open circuit voltage test. A stabilized open circuit voltage reading of 12.6 V or more indicates a fully charged battery.

Answer D is incorrect. Both Technicians are correct.

TASK D.1.19

18. A slack adjuster pushrod is being serviced on a vehicle following a driver tag on a post-trip inspection. Technician A says that the service performed must be logged before the vehicle can be put back into service. Technician B says the service performed should be recorded on a "B" PMI maintenance sheet. Who is correct?

    A. A only

    B. B only

    C. Both A and B

    D. Neither A nor B

Answer A is correct. Only Technician A is correct. Whenever any repair procedure or adjustment is performed on a vehicle, the work must be noted using a sign-off sheet before the vehicle is put back into service. If the problem noted did not require repair, but was still looked at, then this must be recorded as well. A copy of the inspection report must remain with the vehicle.

Answer B is incorrect. A PMI worksheet is not the correct place to record repairs. PMI worksheets are used for work that is scheduled in advance. The following is a list of basic types of PMI inspections: Schedule "A" is a light inspection. Schedule "B" is a more detailed inspection. Schedule "C" is a detailed inspection, service, and adjustment. Schedule "D" is a comprehensive inspection and adjustment.

Answer C is incorrect. Only Technician A is correct.

Answer D is incorrect. Technician A is correct.

TASK C.5

19. What is being performed in the illustration above?

    A. Testing the battery output
    B. Checking the continuity of the ignition switch
    C. Performing a voltage drop on the ground circuit side of the starter
    D. By-passing the magnetic switch

    Answer A is incorrect. A carbon pile and ammeter gauge is the required equipment for testing the output of a battery, not what is shown.

    Answer B is incorrect. An ohmmeter connected in series with the ignition switch is the required equipment and operating procedure.

    Answer C is incorrect. A voltmeter connected to the ground side of the starter circuit is required for a voltage drop test on the starter ground side.

    Answer D is correct. The technician is checking the starter relay operation. The starter relay is a remote mounted magnetic switch that uses low current to control high current flow from the battery to the starter motor. If the starter system has a magnetic switch, jumper around the heavy terminals to see if the motor cranks. If the starter motor cranks, the relay is defective, assuming that control current from the starting switch is available at the small terminal of the relay.

TASK C.3

20. Technician A says an open circuit voltage test may be used in place of a hydrometer test. Technician B says during the open circuit voltage test the headlights should be on. Who is correct?

    A. A only
    B. B only
    C. Both A and B
    D. Neither A nor B

    Answer A is correct. An open circuit voltage test may be used in place of a hydrometer test.

    Answer B is incorrect. During the open circuit voltage test, the headlights should be off.

    Answer C is incorrect. Only Technician A is correct.

    Answer D is incorrect. Technician A is correct.

21. In relation to lubricants (transmission and drive axle), what is channeling?

    A. Lubricant breakdown in high temperatures

    B. Extreme pressures on components causes a lubricant to separate

    C. Liquid seeking the lowest center of gravity

    D. Components starved of lubrication due to cold temperatures

    Answer A is incorrect. Channeling is not a breaking down of a lubricant.

    Answer B is incorrect. Extreme pressures have nothing to do with lubricant separation.

    Answer C is incorrect. Flowing liquids always seek their lowest center of gravity, regardless of location.

    Answer D is correct. Cold-temperature operation can cause transmission and axle lubrication problems known as channeling. High-viscosity oils typically used in transmissions and axles thicken causing greatly reduced flow, or they may not flow at all when the vehicle is first started. Channeling leads to components not getting proper lubrication. Rotating gears can push lubricant aside leaving voids or channels where no lubricant is actually touching the gears.

22. Technician A says that defective and/or improper ground connections are the most common sources of dim lights. Technician B says that when vehicle lighting inspections are performed, it is common practice to start at the back of the trailer and move forward. Who is correct?

    A. A only

    B. B only

    C. Both A and B

    D. Neither A nor B

    Answer A is correct. Defective and/or improper grounds are the most common source of dim lights.

    Answer B is incorrect. Vehicle light inspections typically start at the front of the vehicle.

    Answer C is incorrect. Only Technician A is correct.

    Answer D is incorrect. Technician A is correct.

23. All of the following would cause a constant leak from the air dryer purge valve EXCEPT:

    A. Foreign particles on valve seat.

    B. A purge valve seal damaged.

    C. A purge valve frozen.

    D. A restricted purge line.

    Answer A is incorrect. Foreign particles on the valve seat could cause a constant air leak from the dryer purge valve.

    Answer B is incorrect. A damaged purge valve seat could cause a constant air leak from the dryer purge valve.

    Answer C is incorrect. A frozen purge valve could cause a constant air leak from the dryer purge valve.

    Answer D is correct. A restricted line will not cause a constant air leak from the dryer purge valve; therefore, it is the exception.

24. Technician A says that S-cam-actuated foundation brakes are the most common brake system on heavy-duty trucks today. Technician B says that slack adjusters and S-cams convert linear motion into brake torque. Who is correct?

    TASK D.1.18

    A. A only

    B. B only

    C. Both A and B

    D. Neither A nor B

    Answer A is incorrect. Technician B is also correct.

    Answer B is incorrect. Technician A is also correct.

    Answer C is correct. Both Technicians are correct. S-cam-actuated foundation brake systems are the most common type of foundation brake system found on heavy-duty trucks. Slack adjusters working with S-cams convert the linear force developed by the brake chamber into brake torque.

    Answer D is incorrect. Both Technicians are correct.

25. The diameter on a bake drum is greater at the edges of the friction surfaces than in the center. This drum condition is:

    TASK D.1.16

    A. Bell-mouthed drum.

    B. Concave brake drum.

    C. Drum out-of-round.

    D. Convex brake drum.

    Answer A is incorrect. A bell-mouthed condition results when the drum diameter is greater at the edges of the drum next to the backing plate compared to the edge near the wheel.

    Answer B is incorrect. A concave brake drum results when the drum diameter is greater in the center compared to both edges.

    Answer C is incorrect. An out-of-round condition results when there is a variation in readings 180 degrees apart.

    Answer D is correct. A convex drum results when the diameter on a brake drum is greater at the edges of the friction surface than in the center.

26. A digital diagnostic reader (DDR) is connected to the American Trucking Association (ATA) connector during a routine PMI to read engine data. Technician A says active codes can be read. Technician B says that historic codes can be erased. Who is correct?

    TASK B.7

    A. A only

    B. B only

    C. Both A and B

    D. Neither A nor B

    Answer A is incorrect. Technician B is also correct.

    Answer B is incorrect. Technician A is also correct.

    Answer C is correct. Both Technicians are correct. A DDR can both read active codes and erase historic (inactive) codes.

    Answer D is incorrect. Both Technicians are correct.

2013 © Delmar, Cengage Learning

TASK C.11

27. What does the arrow in the illustration above indicate?

A. Clearance light

B. Identification light

C. Fog light

D. Intermediate light

Answer A is correct. The clearance light is shown and mounted on the extreme right or left side of the cab. Federal lighting regulations require all vehicles 80 inches or wider to have two clearance lamps at the front and two at the rear at the widest point and as high as practicable.

Answer B is incorrect. Identification lights are in the center of the roof of the cab. Federal lighting regulations for the front of vehicles 80 inches or wider require three identification lights spaced 6–12 inches on center as high as practicable.

Answer C is incorrect. Fog lights are usually mounted as close to the ground as possible, usually on the bumper.

Answer D is incorrect. Intermediate lights do not exist.

TASK D.3.3

28. While performing routine maintenance, a technician observes that the power steering system has a filter. Technician A says the filter should be replaced, if required, on a preventive maintenance inspection (PMI) according to the OEM. Technician B says the filter should be replaced after a component failure. Who is correct?

A. A only

B. B only

C. Both A and B

D. Neither A nor B

Answer A is incorrect. Technician B is also correct.

Answer B is incorrect. Technician A is also correct.

Answer C is correct. Both Technicians are correct. Trucks equipped with power steering filters should have them changed, if required, as part of PMIs. When there has been a component failure, the power steering filter may be filled with debris from the failure, and should be changed.

Answer D is incorrect. Both Technicians are correct.

29. All of the following could cause the power-steering-pump pulley to become misaligned EXCEPT:

    A. An over-pressed pulley.

    B. Loose fit from pulley hub to pump shaft.

    C. A loose pump mounting bracket.

    D. A broken engine mount.

TASKS A.4, D.3.2

Answer A is incorrect. An over-pressed pulley could cause the power-steering-pump pulley to become misaligned.

Answer B is incorrect. A loose fit from the pulley hub to pump shaft could cause the power-steering-pump pulley to become misaligned.

Answer C is incorrect. A worn or loose pump mounting bracket could cause the power-steering-pump pulley to become misaligned.

Answer D is correct. A broken engine mount will not cause the power-steering-pump pulley to become misaligned.

30. Technician A says that air horns should be checked during a PMI. Technician B says any physical damage found on the front bumper of a trailer should be noted during a PMI. Who is correct?

    A. A only

    B. B only

    C. Both A and B

    D. Neither A nor B

TASKS B.8, B.27

Answer A is incorrect. Technician B is also correct.

Answer B is incorrect. Technician A is also correct.

Answer C is correct. Both Technicians are correct. Air and electric horns are important safety devices enabling the operator to warn other motorists of imminent danger. Both are required to be fully functional. Any observed problems with a trailer's front bumper, fairings, or mounts (and many other items) should be noted during a PMI.

Answer D is incorrect. Both Technicians are correct.

31. Technician A says that a minor cooling system leak is sufficient criteria for deadlining. Technician B says that the engine should be checked for both static and dynamic leaks. Who is correct?

    A. A only

    B. B only

    C. Both A and B

    D. Neither A nor B

TASK A.5

Answer A is incorrect. A cooling system leak is not considered a safety priority.

Answer B is correct. The engine should be checked for both static and dynamic leaks. This means that the engine should be checked while it is running and while it is off.

Answer C is incorrect. Only Technician B is correct.

Answer D is incorrect. Technician B is correct.

2013 © Delmar, Cengage Learning

TASK D.4.7

32. A fifth wheel is being mounted on a tractor. Technician A says that if it is located too far forward, hard steering and tire wear can result. Technician B says the fifth wheel should be mounted to the frame flanges by welding. Who is correct?

   A. A only
   B. B only
   C. Both A and B
   D. Neither A nor B

   Answer A is correct. If you locate a fifth wheel too far forward on the tractor, hard steering and tire wear can result.

   Answer B is incorrect. You never weld anything to the frame flanges.

   Answer C is incorrect. Only Technician A is correct.

   Answer D is incorrect. Technician A is correct.

TASK D.1.5

33. With the brakes applied at 80 psi (551 kPa), and checking for maximum allowable leakage at the brake application valve exhaust port, all of the following are incorrect EXCEPT:

   A. A 2-inch (50.8 mm) bubble in 1 second.
   B. A 2-inch (50.8 mm) bubble in 2 seconds.
   C. A 3-inch (76.2 mm) bubble in 3 seconds.
   D. A 1-inch (25.4 mm) bubble in 3 seconds.

   Answer A is incorrect. Leakage at the brake application valve exhaust port with the brakes applied should not exceed a 1-inch (2.54 mm) bubble in 3 seconds.

   Answer B is incorrect. Leakage at the brake application valve exhaust port with the brakes applied should not exceed a 1-inch (2.54 mm) bubble in 3 seconds.

   Answer C is incorrect. Leakage at the brake application valve exhaust port with the brakes applied should not exceed a 1-inch (2.54 mm) bubble in 3 seconds.

   Answer D is correct. Leakage at the brake application valve exhaust port with the brakes applied should not exceed a 1-inch (2.54 mm) bubble in 3 seconds.

34. When diagnosing an ABS with an ABS check switch and a light in the instrument panel, all of the following are incorrect EXCEPT:

    A. When the ABS check switch and the ignition switch are turned on, the ABS light flashes blink codes.

    B. The first set of flashes in the blink code sequence indicates whether the system is ABS only or ABS and automatic traction control (ATC).

    C. When there are no electrical defects in the ABS, a 0-0-0 blink code is revealed.

    D. A blink code indicates whether an electrical defect is in a wheel speed sensor or the connecting wiring.

TASK D.1.12

    Answer A is correct. When the ABS check switch and the ignition switch are turned on, the ABS light flashes blink codes.

    Answer B is incorrect. The first set of flashes indicates the ABS configuration.

    Answer C is incorrect. When there are no electrical defects in the ABS, a 1-0-0, 2-0-0, or 4-0-0 blink code is provided.

    Answer D is incorrect. A blink code indicates a defect in a certain area, but not necessarily in a specific component.

35. A customer has a truck towed into the shop and says the starter would not crank the engine over. What should be checked first?

    A. The ground cable connection

    B. The starter solenoid circuit

    C. The ignition switch crank circuit

    D. The battery condition

TASK C.4

    Answer A is incorrect. The battery condition should be checked first.

    Answer B is incorrect. The battery condition should be checked first.

    Answer C is incorrect. The battery condition should be checked first.

    Answer D is correct. The battery condition should be checked first.

36. Technician A says that an engine can be checked for leaks while the engine is off. Technician B says to also check the engine for leaks while it is running. Who is correct?

    A. A only

    B. B only

    C. Both A and B

    D. Neither A nor B

TASK A.5

    Answer A is incorrect. Technician B is also correct.

    Answer B is incorrect. Technician A is also correct.

    Answer C is correct. Both Technicians are correct. An engine should be checked while doing an inspection for both static and dynamic leaks. This means that the engine should be checked while it is running and while it is off. This ensures that all gasket and mating surfaces are monitored in all possible operating and non-operating conditions. When checking for fuel leaks on an engine, it is mandatory that the engine be operating so the fuel system is pressurized, thus making a leak very apparent.

    Answer D is incorrect. Both Technicians are correct.

TASK D.1.14

37. All of the following could cause excessive leakage in an air brake system with the service brakes applied EXCEPT:

    A. A leaking brake chamber diaphragm.

    B. A leaking hose, tube, or fitting.

    C. Defective compressor gaskets.

    D. A bad relay valve.

    Answer A is incorrect. A damaged or defective brake chamber diaphragm may cause excessive leakage.

    Answer B is incorrect. A damaged hose or fitting may cause leakage when pressurized.

    Answer C is correct. Defective compressor gaskets will not cause brake system leakage.

    Answer D is incorrect. A defective relay valve may cause an excessive air leak.

TASK D.2.4

38. Technician A says a correctly adjusted pull type clutch requires ½ inch (1.25 cm) of clearance between the release bearing and the clutch brake disc. Technician B says a correctly adjusted pull type clutch release fork needs ⅛ inch (0.3125 cm) clearance between itself and the release bearing. Who is correct?

    A. A only

    B. B only

    C. Both A and B

    D. Neither A nor B

    Answer A is incorrect. Technician B is also correct.

    Answer B is incorrect. Technician A is also correct.

    Answer C is correct. Both Technicians are correct. The ½-inch measurement between the release bearing and clutch brake provides the necessary pedal travel to produce release bearing to clutch brake contact at the full pedal position. The ⅛-inch release fork to release-bearing clearance produces the correct clutch-pedal freeplay.

    Answer D is incorrect. Both Technicians are correct.

TASK D.2.10

39. When inspecting driveshafts, all of the following can result in vibration EXCEPT:

    A. Missing balance weight.

    B. Out of phase condition.

    C. Light rust on the driveshaft surface.

    D. Loose U-joint cap.

    Answer A is incorrect. Missing driveshaft balance weights will lead to vibration. This type of vibration is directly related to driveshaft RPM and most noticeable at a specific driveshaft speed.

    Answer B is incorrect. An out of phase condition can result in driveshaft vibrations. A driveshaft is in phase when the lugs on the tube yoke and slip yoke line up with each other.

    Answer C is correct. Light rust on the driveshaft surface will not affect driveshaft balance.

    Answer D is incorrect. A loose U-joint cap will cause driveshaft vibration.

40. Technician A says a battery drain of 2 amperes would cause the battery to become discharged. Technician B says the battery drain test should be performed with the engine at normal operating temperature. Who is correct?

TASK C.3

   A. A only
   B. B only
   C. Both A and B
   D. Neither A nor B

   Answer A is correct. A battery drain of 2 amperes would cause the battery to become discharged.

   Answer B is incorrect. A battery drain test may be performed at any engine temperature.

   Answer C is incorrect. Only Technician A is correct.

   Answer D is incorrect. Technician A is correct.

41. During an inspection, a tractor must be placed out of service if any of these tire conditions are present EXCEPT:

TASKS D.3.10, D.3.13

   A. If a tire is mounted or inflated so it contacts any part of the vehicle.
   B. If a drive axle tire has a tread depth of less than 1/32 inch at two adjacent tread grooves at three separate locations.
   C. If a steer axle tire has a tread depth of 5/32 inch at two adjacent tread grooves at any location on the tire.
   D. If the tire is labeled "Not for Highway Use".

   Answer A is incorrect. If a tire is mounted or inflated so it contacts any part of the vehicle, this problem must be corrected before placing the vehicle in service.

   Answer B is incorrect. If a drive axle tire has a tread depth of less than 1/32 inch at two adjacent tread grooves at three separate locations, this problem must be corrected before placing the vehicle in service.

   Answer C is correct. If a steer axle tire has a tread depth of 5/32 inch at two adjacent tread grooves at any location on the tire, the tire is satisfactory. Minimum tread depth is 4/32 inch.

   Answer D is incorrect. If the tire is labeled "Not for Highway Use," this problem must be corrected before placing the vehicle in service.

42. An electronic cruise control maintains the vehicle speed at 45 mph (72 km/h) when the selected speed is 60 mph (96 km/h). What is the most likely cause of this problem?

TASK E.6

   A. A defective cruise control module
   B. A defective vehicle speed sensor (VSS)
   C. Improper size of tires on the drive axles
   D. A defective cruise control switch

   Answer A is incorrect. A defective cruise control module is not the most likely cause of improper cruise set speed.

   Answer B is correct. A VSS is the most likely cause of improper cruise control set speed.

   Answer C is incorrect. Improper size of tires on the drive axles is not the most likely cause of improper cruise control set speed.

   Answer D is incorrect. A defective cruise control switch is not the most likely cause of improper cruise control set speed.

TASK D.3.11

43. During a tire inspection, a technician notices a cut in the sidewall of a tire. Which of the following is the most likely cause?

    A. Excessive toe-in

    B. Camber

    C. Careless driving

    D. Caster

Answer A is incorrect. Excessive toe-in will not cause a cut in the sidewall. Improper toe adjustment causes a feathered tire tread wear pattern.

Answer B is incorrect. Camber maladjustment will not cause a cut in the sidewall. Improper camber causes wear on one side of the tire tread.

Answer C is correct. Careless driving is the most likely cause of a cut in the sidewall. Sidewall damage can lead to future tire failure.

Answer D is incorrect. Caster will not cause a cut in the sidewall.

TASK D.4.15

44. Technician A says that impact guards are required to be equipped on vehicles manufactured after 1998. Technician B says that the impact guards should be carefully inspected for damage during routine PMI inspection since they are subject to abuse. Who is correct?

    A. A only

    B. B only

    C. Both A and B

    D. Neither A nor B

Answer A is incorrect. Technician B is also correct.

Answer B is incorrect. Technician A is also correct.

Answer C is correct. Both Technicians are correct. Vehicles (straight trucks and semi-trailers) manufactured after 1998 are required to be equipped with collapsible docking guards. The guards should be examined for damage during a PMI inspection since they are subject to constant abuse. Check the guards and mounting hardware.

Answer D is incorrect. Both Technicians are correct.

TASKS
A.5, A.15

45. A turbocharger is being inspected for oil leaks. Technician A says to check it while the engine is running. Technician B measures turbine shaft radial runout with the engine running. Who is correct?

    A. A only

    B. B only

    C. Both A and B

    D. Neither A nor B

Answer A is correct. Only Technician A is correct. A turbocharger should be checked for oil leaks when the engine is running. Without the engine producing oil pressure, it would be more difficult to detect leaks. The oil pressure to a turbocharger feeds the bearings, spills to an oil drain cavity, and then drains to the crankcase by means of a return hose.

Answer B is incorrect. It would be dangerous and impossible to try to check the turbocharger shaft for runout while the engine is running. When running a turbocharger-equipped diesel engine without the air intake, be sure to install an intake guard over the turbocharger inlet to avoid personal injury from the exposed rotating impeller.

Answer C is incorrect. Only Technician A is correct.

Answer D is incorrect. Technician A is correct.

46. All of the following are underside components of a trailer EXCEPT:

    A. Bulkhead.

    B. Tires.

    C. Suspension.

    D. Landing gear.

TASKS D.4.1, D.4.2

Answer A is correct. The bulkhead is a trailer body component.

Answer B is incorrect. Tires are a trailer underside component.

Answer C is incorrect. The suspension system is a trailer underside component.

Answer D is incorrect. The landing gear is a trailer underside component.

47. An exhaust brake is being inspected on a truck. Technician A says that this should be performed during the road test. Technician B says that a properly functioning engine compression brake should be noiseless. Who is correct?

    A. A only

    B. B only

    C. Both A and B

    D. Neither A nor B

TASK E.9

Answer A is correct. Only Technician A is correct. The engine/exhaust brake should be tested on a road test. Use the engine brake during the road test when the engine RPM is in torque rise. Note the disengage RPM.

Answer B is incorrect. All engine compression brakes produce exhaust noise. When the compression brake is energized, it causes the engine to perform like a power-absorbing air compressor. The compression brake system uses engine oil pressure to open the exhaust valves before the compression stroke is complete and combustion does not occur in the cylinder. The compressed air is released into the engine exhaust system.

Answer C is incorrect. Only Technician A is correct.

Answer D is incorrect. Technician A is correct.

48. A cooling system on a heavy truck is being tested. Technician A says that the color of the coolant indicates whether it is EG, PG, or ELC. Technician B says to drain off a gallon of coolant before performing a pressure test. Who is correct?

    A. A only

    B. B only

    C. Both A and B

    D. Neither A nor B

TASKS A.20, A.23

Answer A is incorrect. Coolant is not color-coded by type.

Answer B is incorrect. You only check the cooling system at full capacity.

Answer C is incorrect. Neither Technician is correct.

Answer D is correct. Neither Technician is correct. It is important to know that all coolants are not alike and are not always compatible with each other. Identify the coolant being used and be sure to follow the pertinent change intervals and refill with the correct type. Extended life coolants (ELCs) have a longer service life than ethylene glycol (EG) and propylene glycol (PG) coolants. It is important to test the radiator cap to ensure that it will retain the indicated pressure setting. Using a pressure tester unit, install the cap onto the tester and apply pressure to the specified pressure release point. If the specified limit cannot be achieved, then cap replacement is required. Pressure testing the engine cooling system for leaks is performed in the same manner. Apply pressure by means of the same tester unit. The gauge reading should achieve the same setting as the recommended radiator cap. The gauge reading should hold steady.

TASK C.12

49. Which of the following would be the least important part of a trailer's lighting system to check?

    A. Side amber marker light(s)

    B. Interior overhead light(s)

    C. Main wiring input socket and associated wiring

    D. Reefer power supply

Answer A is incorrect. Side amber marker lights are a mandatory check. Check for burned-out, damaged, or missing marker lights and reflectors.

Answer B is incorrect. All lighting, both interior and exterior, is to be inspected. Check for: proper operation, color and cleanliness of headlights, turn signals, stoplights, taillights, emergency flashers, reflectors, marker lights, clearance lights, cab instrumentation and interior lights, and trailer lighting.

Answer C is incorrect. All wiring connectors are checked when performing an inspection. Wiring connectors should be clean and tight. Wires and/or harnesses must be replaced if insulation becomes burned, cracked, or deteriorated.

Answer D is correct. The reefer power supply is not part of the trailer's lighting system.

TASKS D.3.16, D.3.18

50. Technician A says that re-torquing wheel lug nuts are not required when performing a PMI. Technician B says that increased load carrying capacity is an advantage of switching to low-profile tires. Who is correct?

    A. A only

    B. B only

    C. Both A and B

    D. Neither A nor B

Answer A is incorrect. Re-torquing of wheel lug nuts is required when performing a PMI. Answer B is incorrect. Increased load-carrying capacity is not an advantage of switching to low-profile tires.

Answer C is incorrect. Neither Technician is correct.

Answer D is correct. Neither Technician is correct. At recommended intervals re-torquing is necessary. This is due to the constant flexing of the stud/hub material. Overloading can be a factor. Manufacturers have specified intervals recommended for wheel and rim inspections. Low-profile tires decrease the load-carrying capacity of a vehicle. The low-profile tire does offer several advantages compared to conventional tires, which include reduced rolling resistance, cooler running, longer tread life, improved fuel economy, reduced vehicle height with a lower center of gravity, and improved traction.

PREPARATION EXAM 5—ANSWER KEY

1.	A	18.	A	35.	B
2.	B	19.	C	36.	B
3.	C	20.	D	37.	C
4.	C	21.	A	38.	C
5.	A	22.	D	39.	A
6.	C	23.	D	40.	C
7.	A	24.	B	41.	D
8.	B	25.	A	42.	C
9.	D	26.	C	43.	B
10.	D	27.	A	44.	B
11.	A	28.	C	45.	C
12.	C	29.	C	46.	B
13.	B	30.	B	47.	C
14.	B	31.	B	48.	C
15.	C	32.	C	49.	B
16.	A	33.	C	50.	A
17.	B	34.	D		

PREPARATION EXAM 5—EXPLANATIONS

1. Technician A says that a fully charged battery will not freeze as easily as a discharged battery. Technician B says that the specific charge of a battery has no effect on the electrolyte freezing. Who is correct?

    TASK C.3

    A. A only
    B. B only
    C. Both A and B
    D. Neither A nor B

    Answer A is correct. Only Technician A is correct. A fully charged battery will not freeze as easily as a discharged battery.

    Answer B is incorrect. A battery that is even partially discharged will be exposed to freezing problems. The specific gravity in a discharged battery is decreased to a point where the electrolyte can freeze.

    Answer C is incorrect. Only Technician A is correct.

    Answer D is incorrect. Technician A is correct.

TASKS E.5,
E.7, E.9

2. A truck is being road tested as part of a scheduled preventive maintenance inspection (PMI). All of the following should be checked EXCEPT:

A. Check road speed limiter.

B. Perform snap-shot test for active codes.

C. Observe exhaust for excessive smoke.

D. Verify engine brake operation.

Answer A is incorrect. Checking the road speed limiter is part of a vehicle road test. Connect the appropriate engine scan tool, OEM special tool, or laptop computer to the truck's diagnostic connector. During the road test, activate the tool's parameter specification readout and record the road speed. Reprogram if required.

Answer B is correct. A snapshot test for active codes is a diagnostic procedure that would not be performed as part of a normal scheduled PMI. A snapshot test readout from the ECM can be used to facilitate troubleshooting intermittent problems that either do not generate codes or log a code with no clear reason.

Answer C is incorrect. Observing exhaust for excessive smoke is part of a road test. The color and consistency of the exhaust can reveal combustion problems. Oil entering the combustion chambers causes blue smoke in the exhaust. Unburned fuel, insufficient intake air, or an improper grade of fuel produces a gray or black smoke. Water or coolant entering the combustion chambers produces a white steam in the exhaust. White smoke may also indicate that injectors are misfiring. White smoke also occurs either when the engine starts the first time or when the ambient temperature is too low for proper combustion.

Answer D is incorrect. Verification of proper engine brake operation is performed during a road test. Use the engine brake during the road test when the engine RPM is in torque rise. Note the disengage RPM.

TASKS D.2.12,
D.2.13

3. Technician A says that lubricant leaking from the drive axle seals could be caused by a plugged breather. Technician B says that grease fittings should be wiped before lubricating. Who is correct?

A. A only

B. B only

C. Both A and B

D. Neither A nor B

Answer A is incorrect. Technician B is also correct.

Answer B is incorrect. Technician A is also correct.

Answer C is correct. Both Technicians are correct. A plugged breather can cause a pressure build-up in the axle, forcing fluid past the seals and gaskets. Carefully investigate any source leaks found on the axle differential. Replacing the seal if the axle differential breather is plugged will not cure leaking seals and the problem will continue to exist. Grease fittings should be cleaned off before lubricating to remove accumulated grease and abrasives. Otherwise, dirt and contaminants will be forced into the component being lubricated. While supplying grease through the grease fittings, check to be sure they are not missing, damaged, broken, or plugged.

Answer D is incorrect. Both Technicians are correct.

4. When checking the coolant condition with a supplemental coolant additive (SCA) test strip, the technician finds that the coolant pH condition is higher than specification limits. While reviewing maintenance records the technician finds that the pH condition was higher than the limit at the last service also. Which of these items should the technician do?

TASK A.24

   A. Add more antifreeze to increase SCA.

   B. Continue to run the truck until the next PMI.

   C. Drain the entire cooling system and add the proper SCA mixture.

   D. Run the truck with no SCA additives until the next PMI.

   Answer A is incorrect. SCA stands for supplemental coolant additive, which is a corrosion inhibitor additive. Adding more antifreeze does not change SCA reading.

   Answer B is incorrect. Continuing to run the truck until the next PMI will not change the SCA reading.

   Answer C is correct. Drain the entire cooling system and add the proper SCA mixture to achieve the correct percentage of SCA.

   Answer D is incorrect. This will not change the SCA reading.

2013 © Delmar, Cengage Learning

5. Which of the following would not be part of performing a PMI on a truck with a fifth wheel, like the one shown in the illustration above?

TASKS D.4.7,
D.4.8

   A. Reverse the location of the jaws every 100,000 miles.

   B. Make sure horizontal movement does not exceed ⅛ inch.

   C. Torque mounting bolts.

   D. Inspect the locking mechanism.

   Answer A is correct. The jaws are not reversible. The jaws should be inspected and replaced if worn.

   Answer B is incorrect. The fifth wheel should be checked for horizontal movement during a PMI. When checking fifth-wheel adjustment, movement should not exceed ⅛-inch horizontal play at any time.

   Answer C is incorrect. Fastener integrity and torque should be checked at each PMI.

   Answer D is incorrect. Locking mechanism should be inspected during a PMI. The locking mechanism may be released mechanically with a lever or by air pressure supplied to an air cylinder. Test the operation of either type during the PMI.

**TASKS C.2,
C.5, C.6**

6. What would be the LEAST LIKELY cause of a high-amperage reading when checking starter current draw?

　　A.　Short in the starter motor

　　B.　Mechanical resistance due to binding engine

　　C.　Starter wire gauge is too small

　　D.　Short to ground in the starter wire

Answer A is incorrect. A short to ground through the starter motor would cause a high current draw.

Answer B is incorrect. Any mechanical resistance would require more energy, thus more current.

Answer C is correct. Small gauge starter wire will cause high resistance and low current draw.

Answer D is incorrect. A short to ground in the starter wire would cause increased current flow.

TASK D.1.12

7. Technician A says that all electronic ABS are equipped with self-diagnostics. Technician B says that ABS is an option on new trucks. Who is correct?

　　A.　A only

　　B.　B only

　　C.　Both A and B

　　D.　Neither A nor B

Answer A is correct. Only Technician A is correct. All electronic ABS are equipped with self-diagnostics. These can be accessed by flash or blink codes using the dash warning light or a scan tool (digital diagnostic reader).

Answer B is incorrect. New trucks are equipped with ABS brakes and are mandatory.

Answer C is incorrect. Only Technician A is correct.

Answer D is incorrect. Technician A is correct.

TASK A.5

8. Technician A says that a fuel leak while the engine is running is less hazardous than when the engine is off. Technician B says that the engine should be checked for both static and dynamic leaks. Who is correct?

　　A.　A only

　　B.　B only

　　C.　Both A and B

　　D.　Neither A nor B

Answer A is incorrect. A fuel leak is more hazardous while the engine is running than when the engine is off.

Answer B is correct. Only Technician B is correct. The engine should be checked for both static and dynamic leaks. This means that the engine should be checked while it is running and while it is off.

Answer C is incorrect. Only Technician B is correct.

Answer D is incorrect. Technician B is correct.

9. Which of the following is the LEAST LIKELY cause of water droplets or gray oil on the dipstick?

    A. Cracked block or cylinder head

    B. Blown head gasket

    C. Leaky oil cooler

    D. Worn oil pump

TASK A.7

Answer A is incorrect. The indication is coolant in the oil, and a cracked block or cylinder head causes this condition.

Answer B is incorrect. The indication is coolant in the oil, and a blown head gasket causes this condition.

Answer C is incorrect. The indication is water in the oil, and a leaky oil cooler causes this condition.

Answer D is correct. A worn oil pump is the least likely cause of water droplets or gray oil on the dipstick because there is no connection between the cooling system and the oil pump.

10. The major changes introduced by the Clean Air Act in heavy-duty truck A/C systems relate to:

    A. Certification of A/C technicians.

    B. Maintaining older equipment.

    C. Replacing leaking equipment.

    D. Refrigerants and their handling.

TASKS B.28, B.29

Answer A is incorrect. Technician certification is not covered under the Clean Air Act.

Answer B is incorrect. Standard service and preventive maintenance inspection (PMI) records cover this area.

Answer C is incorrect. This is part of a PMI.

Answer D is correct. Refrigerants and their handling are mandated by the Federal Clean Air Act Amendment.

11. Technician A says that the tractor protection valve protects the tractor air supply under trailer breakaway conditions. Technician B says that the trailer supply dash valve (emergency) is diamond shaped and yellow in color. Who is correct?

    A. A only

    B. B only

    C. Both A and B

    D. Neither A nor B

TASKS D.1.1, D.1.9

Answer A is correct. Only Technician A is correct. The tractor protection valve protects the tractor air supply under trailer breakaway conditions or severe air leakage.

Answer B is incorrect. The trailer supply dash valve is a red hexagonal valve. The park brake dash valve is a yellow colored diamond shaped valve.

Answer C is incorrect. Only Technician A is correct.

Answer D is incorrect. Technician A is correct.

TASK B.30

12. A small leak is found in an A/C system. What should the technician do?

 A. Nothing. Wait for a service report to be filed.

 B. Log the problem on a PMI schedule sheet.

 C. Log the problem, and then inform the owner/driver that this is in violation of the Clean Air Act.

 D. Repair the leak immediately.

Answer A is incorrect. A technician should never ignore a problem when a potential repair(s) poses a problem to the serviceability of the truck.

Answer B is incorrect. Logging in an A/C serviceable problem is not the minimum responsibility of a technician.

Answer C is correct. The technician logs the problem, and then informs the owner/driver that this is in violation of the Clean Air Act.

Answer D is incorrect. A vehicle should never be repaired without proper paperwork in order.

TASK B.22

13. Technician A says the Department of Transportation (DOT) does not require PMI records on lubrication. Technician B says that maintenance records should be referenced before undertaking major repairs. Who is correct?

 A. A only

 B. B only

 C. Both A and B

 D. Neither A nor B

Answer A is incorrect. PMI records on heavy trucks are required by the DOT.

Answer B is correct. Only Technician B is correct. Maintenance records should be referenced before undertaking major repairs.

Answer C is incorrect. Only Technician B is correct.

Answer D is incorrect. Technician B is correct.

TASK D.1.17

14. Technician A says that the link between the brake chamber and the foundation brake is known as the brake shoe roller. Technician B says that the link between the brake chamber and the foundation brake is the slack adjuster. Who is correct?

 A. A only

 B. B only

 C. Both A and B

 D. Neither A nor B

Answer A is incorrect. The brake shoe roller is part of the foundation brake assembly rides on the S-camshaft.

Answer B is correct. Only Technician B is correct. The slack adjuster is the link between the brake chamber and the foundation brake assembly.

Answer C is incorrect. Only Technician B is correct.

Answer D is incorrect. Technician B is correct.

15. Technician A states that a vehicle should be deadlined if weld repairs are found to have been made on pintle chains and hooks. Technician B says that at least one functioning amber or red light on rear loads projecting more than 4 feet beyond the vehicle body is required. Who is correct?

TASK B.9

 A. A only

 B. B only

 C. Both A and B

 D. Neither A nor B

Answer A is incorrect. Technician B is also correct.

Answer B is incorrect. Technician A is also correct.

Answer C is correct. Both Technicians are correct. The vehicle should be deadlined if weld repairs are found to be made on pintle chains and hooks. CVSA vehicle out-of-service criteria include cracks anywhere in the pintle hook assembly or any visible welded repairs to the pintle hook. Out-of-service criteria for pintle chains and hooks also lists improper repairs to chains and hooks including welding, wire, small bolts, rope, and tape. Loads projecting more than 4 feet beyond the vehicle body must have at least one operative red or amber light on the rear of loads visible from 500 feet.

Answer D is incorrect. Both Technicians are correct.

2013 © Delmar, Cengage Learning

16. In the illustration above, what could commonly be mounted above the roof depicted by arrow?

TASK C.11

 A. Identification lights

 B. Solar panels

 C. Skylight

 D. Air vents for sleeper cab

Answer A is correct. Identification lights are typically roof-mounted in this spot.

Answer B is incorrect. Roof-mounted solar panels are not commonly used on trucks.

Answer C is incorrect. Skylights are not generally used on a heavy truck.

Answer D is incorrect. Air vents are not as common on the roof as clearance lights.

TASK A.9

17. Technician A says that fuel tanks should be left near empty after long runs before refilling. Technician B says that after a long run, the fuel tanks should be refilled and not left empty. Who is correct?

A. A only

B. B only

C. Both A and B

D. Neither A nor B

Answer A is incorrect. The fuel tanks should be refilled after a long run.

Answer B is correct. Only Technician B is correct. The fuel tanks should be refilled after a long run. Moisture that is present in the air will condense in a near empty tank.

Answer C is incorrect. Only Technician B is correct.

Answer D is incorrect. Technician B is correct.

TASK A.15

18. Technician A says that turbochargers perform at very high speeds which can be in excess of 100,000 rpm. Technician B says that turbochargers only turn at engine RPM. Who is correct?

A. A only

B. B only

C. Both A and B

D. Neither A nor B

Answer A is correct. Only Technician A is correct. Modern diesel engine turbochargers perform at very high speeds which can be in excess of 100,000 rpm.

Answer B is incorrect. Modern diesel engine turbochargers perform at very high speeds which can be in excess of 100,000 rpm.

Answer C is incorrect. Only Technician A is correct.

Answer D is incorrect. Technician A is correct.

TASKS C.8, C.9

19. Technician A says that the condition of the batteries should be verified with a load test prior to performing an alternator current output test. Technician B says that a load tester with an inductive pick-up may be used to test alternator current output. Who is correct?

A. A only

B. B only

C. Both A and B

D. Neither A nor B

Answer A is incorrect. Technician B is also correct.

Answer B is incorrect. Technician A is also correct.

Answer C is correct. Both Technicians are correct. A carbon pile load tester can be used to perform an alternator output voltage and amperage test. Connect it across the battery, observing correct polarity. Start the engine and run at about 1,500 rpm. Observe the unloaded output voltage. It should be between 13.5 to 14.5 volts. Attach the inductive pick-up to the alternator output wire. With the engine at high idle, use the load tester to draw the highest amperage reading without dropping the voltage to fall below 12 volts. This procedure forces the alternator to increase its output to near its maximum. The ammeter should indicate output within 5 percent of the alternator's capacity.

Answer D is incorrect. Both Technicians are correct.

20. Technician A says that when checking brake chamber stroke using the applied stroke method to use 25 to 35 psi. Technician B says that when checking brake chamber stroke using the applied stroke method to use only 10 psi. Who is correct?

TASK D.1.17

   A. A only

   B. B only

   C. Both A and B

   D. Neither A nor B

   Answer A is incorrect. Using only 25 to 35 psi for checking brake chamber stroke using the applied stroke measurement will result in an inaccurate reading. Applied stroke method uses 80 to 90 psi.

   Answer B is incorrect. Using only 10 psi for checking brake chamber stroke using the applied stroke measurement will result in an inaccurate reading. Applied stroke method uses 80 to 90 psi.

   Answer C is incorrect. Neither Technician is correct.

   Answer D is correct. Neither Technician is correct. Brake chamber applied stroke measurement is done using 80 to 90 psi.

21. What is the approximate stoichiometric air-fuel ratio?

TASK A.16

   A. 15:1

   B. 8:1

   C. 10:1

   D. 17.5:1

   Answer A is correct. Air-fuel ratio by mass is approximately 15:1 and is referred to as stoichiometric ratio.

   Answer B is incorrect. Air-fuel ratio by mass is approximately 15:1 and is referred to as stoichiometric ratio.

   Answer C is incorrect. Air-fuel ratio by mass is approximately 15:1 and is referred to as stoichiometric ratio.

   Answer D is incorrect. Air-fuel ratio by mass is approximately 15:1 and is referred to as stoichiometric ratio.

22. Technician A says that a driver must receive a visible alert when system air pressure drops below 100 psi. Technician B says that to check for proper functioning of the inversion (spring brake valve) valve, the air system should be at 60 psi. Who is correct?

TASKS D.1.7, D.1.8

   A. A only

   B. B only

   C. Both A and B

   D. Neither A nor B

   Answer A is incorrect. FMVSS 121 requires that a driver receive a visible alert when system air pressure drops below 60 psi.

   Answer B is incorrect. To check for proper functioning of the inversion valve (spring brake valve), the air system should be fully charged (minimum 100 psi).

   Answer C is incorrect. Neither Technician is correct.

   Answer D is correct. Neither Technician is correct. A low air pressure warning device is fitted to both the primary and secondary circuits. This is a simple electrical switch that can be plumbed anywhere into a system requiring monitoring. The switch is electrically closed whenever the air pressure being monitored is below 60 psi. When the air pressure value exceeds 60 psi, the switch opens. Remember, an inversion valve is a failsafe device and does not operate when brakes are functioning properly.

TASK B.30

23. The driver complains about a rotten egg smell in the truck cab that is most noticeable when the A/C system is first turned on. What could be the cause of this problem?

    A. Leaking evaporator core

    B. Leaking condenser core

    C. Leaking heater core

    D. Plugged evaporator case drain

Answer A is incorrect. A rotten egg smell in the cab is caused by a plugged evaporator case drain.

Answer B is incorrect. A rotten egg smell in the cab is caused by a plugged evaporator case drain.

Answer C is incorrect. A rotten egg smell in the cab is caused by a plugged evaporator case drain.

Answer D is correct. A rotten egg smell in the cab is caused by a plugged evaporator case drain.

2013 © Delmar, Cengage Learning

TASK D.3.6

24. What is the procedure being performed in the illustration above?

    A. Knuckle vertical play

    B. Knuckle upper bushing play measurement

    C. Lower bushing deflection

    D. Kingpin axial play

Answer A is incorrect. Vertical play is checked with the indicator on top of the knuckle.

Answer B is correct. A knuckle upper bushing play measurement is shown in the illustration.

Answer C is incorrect. The indicator should be mounted on the lower bushing area.

Answer D is incorrect. There is no kingpin axial play test.

TASK D.4.6

25. All of these procedures should be done when replacing a torque rod on a tandem axle suspension EXCEPT:

    A. Replace alignment washers with new ones.

    B. Secure the truck's parking brakes and chock the tires.

    C. Ensure the replacement torque rod is of the correct length.

    D. Check the axle alignment when finished.

Answer A is correct. It is not necessary to replace the axle alignment washers when replacing a radius rod on a truck. If the washers are not worn beyond use, then they should be reused to ensure a correct realignment of the axle. Also, the number of washers and their respective positions (forward or rearward) should be noted and returned exactly as removed.

Answer B is incorrect. Securing the vehicle is one of the first operations performed.

Answer C is incorrect. Replacement rods come in many different sizes.

Answer D is incorrect. An axle alignment check is always in order when replacing a torque rod.

26. Technician A says some manufacturers recommend the secondary fuel filter be installed dry and primed with a hand pump. Technician B says that most fuel filters are changed at routine intervals (PMI schedules). Who is correct?

TASK A.10

   A. A only
   B. B only
   C. Both A and B
   D. Neither A nor B

Answer A is incorrect. Technician B is also correct.

Answer B is incorrect. Technician A is also correct.

Answer C is correct. Both Technicians are correct. Most fuel filters are changed at routine intervals (PMI schedules). These filters must be primed with fuel before the engine can be started. Most fuel filters can be filled with clean fuel and reinstalled, but some manufacturers recommend that the secondary filter be installed dry and primed with a hand pump. This procedure ensures that only filtered diesel fuel enters the secondary filter. When hand priming is necessary, a line fitting upstream from the pump should be cracked open and priming should continue until no air bubbles are present at the fitting.

Answer D is incorrect. Both Technicians are correct.

27. Which of the following is the best description of an acceptable axle lubricant level?

TASK D.2.14

   A. Lube level is even with the bottom of the filler hole
   B. Lube level can be easily seen through the filler hole
   C. Lube can be touched one knuckle joint below the bottom of the filler hole
   D. Lube can be touched with the index finger through the filler hole

Answer A is correct. Axle lubrication level is correct only when the fluid level is even with the bottom of the filler hole in the carrier housing with the truck on level ground. The angle of the drive pinion, as mounted under the vehicle, determines which oil fill/level plug is to be used if two are present. Inadequate lubrication from low lube levels could create friction, cause overheating, break down the protective film, and finally result in seizure or welding of mating components. The other choices could indicate a low fluid level.

Answer B is incorrect. Lube level must be even with the bottom of the filler hole.

Answer C is incorrect. Lube level must be even with the bottom of the filler hole.

Answer D is incorrect. Lube level must be even with the bottom of the filler hole.

28. Technician A says it is good practice to check underground fuel storage tanks for water content. Technician B says batteries can be kept from freezing if they have a full charge. Who is correct?

TASKS A.9, C.3

   A. A only
   B. B only
   C. Both A and B
   D. Neither A nor B

Answer A is incorrect. Technician B is also correct.

Answer B is incorrect. Technician A is also correct.

Answer C is correct. Both Technicians are correct. Underground fuel storage tanks can form condensation and should always be checked for water content. Checks need to be diligent and any water found in the fuel system must be removed. A battery that is even partially discharged will be exposed to a freeze-up problem in cold weather. The specific gravity in a discharged battery is decreased to a point where the electrolyte can freeze. Electrolyte in a fully charged battery has a specific gravity of 1.265 and will not freeze until it is approximately below 0 °F. A discharged battery with an electrolyte specific gravity of 1.1 will freeze at approximately 15 °F.

Answer D is incorrect. Both Technicians are correct.

29. A worn front steering bushing can cause all of the following EXCEPT:

    A. A loose feeling in the steering wheel when driving straight.
    B. Noise when hitting a bump in the road.
    C. Excessive use of power steering fluid.
    D. Pull to the left while driving.

    Answer A is incorrect. A loose feeling in steering wheel when driving straight can be caused by a worn front steering bushing.

    Answer B is incorrect. A noise when hitting a bump in the road can be caused by a worn front steering bushing.

    Answer C is correct. Excessive use of power steering fluid is the exception and cannot be caused by a worn steering bushing.

    Answer D is incorrect. A pull to the left while driving on the road can be caused by a worn front steering bushing.

30. Technician A says that if the temperature of a wire increases, its resistance will decrease. Technician B says that if the temperature of a wire increases, its resistance will increase. Who is correct?

    A. A only
    B. B only
    C. Both A and B
    D. Neither A nor B

    Answer A is incorrect. As the temperature of a wire increases, its resistance will increase.

    Answer B is correct. Only Technician B is correct. As the temperature of a wire increases, its resistance will increase.

    Answer C is incorrect. Only Technician B is correct.

    Answer D is incorrect. Technician B is correct.

31. What is the most common cause of brake imbalance on a truck or trailer?

    A. Different size brake chambers
    B. Brakes out of adjustment
    C. Mismatched brake linings
    D. Mismatched brake hose sizes

    Answer A is incorrect. Though different size brake chambers can cause brake imbalance, out of adjustment brakes are the most common cause of brake imbalance.

    Answer B is correct. Brakes out of adjustment are the most common cause of brake imbalance.

    Answer C is incorrect. Though mismatched brake linings can cause brake imbalance, out of adjustment brakes are the most common cause of brake imbalance.

    Answer D is incorrect. Though mismatched brake hose sizes can cause brake imbalance, out of adjustment brakes are the most common cause of brake imbalance.

32. A truck has a slower than normal air pressure buildup time. Technician A says that leaking air lines, connections, or valves could be the cause. Technician B says a worn air compressor could be the cause. Who is correct?

TASK D.1.5

A. A only

B. B only

C. Both A and B

D. Neither A nor B

Answer A is incorrect. Technician B is also correct.

Answer B is incorrect. Technician A is also correct.

Answer C is correct. Leaking air lines, connections, or valves can be the cause of slower than normal air pressure buildup times. A worn air compressor can also contribute to slower than normal air pressure buildup times.

Answer D is incorrect. Both Technicians are correct.

33. Technician A says a circuit breaker is designed to protect a circuit from high current. Technician B says a circuit breaker's rating is given in amps. Who is correct?

TASK C.10

A. A only

B. B only

C. Both A and B

D. Neither A nor B

Answer A is incorrect. Technician B is also correct.

Answer B is incorrect. Technician A is also correct.

Answer C is correct. Both Technicians are correct. Circuit breakers are designed to protect circuits from high current, and are rated in amps according to circuit protection needs.

Answer D is incorrect. Both Technicians are correct.

34. Which of the following should be done when performing a preventive maintenance inspection (PMI) on a heavy truck fuel system?

TASK A.10

A. Check the pour point of the fuel.

B. Verify the accuracy of the fuel gauge.

C. Check the fuel pump for volumetric capacity.

D. Replace the fuel filter and prime the system.

Answer A is incorrect. A pour point check of the fuel is not done during a PMI.

Answer B is incorrect. Dash gauge checks are part of an electrical system check.

Answer C is incorrect. A test of the pump is only performed when there is a performance problem.

Answer D is correct. Replacing the fuel filter and priming the system is done during a PMI. Filters must be primed with fuel before the engine can be started. Most fuel filters can be filled with clean fuel and reinstalled, but some manufacturers recommend that the secondary filter be installed dry and primed with a hand pump. This procedure ensures that only filtered diesel fuel enters the secondary filter. Fuel contamination can occur easily if the filter is primed with unfiltered fuel or dirt is allowed to enter when the element is being changed.

**TASKS C.5,
C.8, C.10**

35. Technician A says that a circuit should be energized when using an ohmmeter. Technician B says that a voltage drop test is an accurate means of identifying high resistance in an energized circuit. Who is correct?

   A.   A only
   B.   B only
   C.   Both A and B
   D.   Neither A nor B

   Answer A is incorrect. Performing a continuity test (ohms) involves using the power source of the meter to determine resistance. Source voltage should be disconnected before using an ohmmeter.

   Answer B is correct. Only Technician B is correct. Using voltage drop is an accurate means of identifying high resistance in an energized circuit.

   Answer C is incorrect. Only Technician B is correct.

   Answer D is incorrect. Technician B is correct.

TASK D.1.6

36. Technician A says that the primary tank feeds the supply tanks in an air brake system. Technician B says that the supply tank feeds the primary and secondary tanks in an air brake system. Who is correct?

   A.   A only
   B.   B only
   C.   Both A and B
   D.   Neither A nor B

   Answer A is incorrect. Only Technician B is correct. The supply tank feeds the primary and secondary tanks in an air brake system.

   Answer B is correct. The supply tank feeds the primary and secondary tanks in an air brake system.

   Answer C is incorrect. Only Technician B is correct.

   Answer D is incorrect. Technician B is correct.

TASK C.10

37. Technician A says that an open circuit can be caused by an open switch. Technician B says that an open circuit can be caused by a broken wire. Who is correct?

   A.   A only
   B.   B only
   C.   Both A and B
   D.   Neither A nor B

   Answer A is incorrect. Technician B is also correct.

   Answer B is incorrect. Technician A is also correct.

   Answer C is correct. Both Technicians are correct. An open circuit can be caused by open switches, broken wires, bad connections, and blown fuses.

   Answer D is incorrect. Both Technicians are correct.

38. What is the best way to clean a rear-axle breather?

    A. Replace the breather.

    B. Soak the breather in gasoline.

    C. Soak in solvent, and then blow out with compressed air.

    D. Use a rag to wipe the vent orifice clean.

TASK D.2.12

    Answer A is incorrect. Replacing the breather is not cleaning the breather.

    Answer B is incorrect. Soaking the breather in gasoline is not a recommended procedure.

    Answer C is correct. The best way to clean a rear-axle breather is to soak in solvent and then use compressed air.

    Answer D is incorrect. Wiping the orifice will not clean any dirt trapped inside the breather.

39. Technician A says that when considering whether to deadline a vehicle, it is necessary to weigh the importance of getting a load moved versus a mechanical problem that could cause a breakdown but is not safety related. Technician B says that a vehicle should be deadlined for a cruise control malfunction. Who is correct?

    A. A only

    B. B only

    C. Both A and B

    D. Neither A nor B

TASKS A.1, E.6

    Answer A is correct. Only Technician A is correct. When considering whether to deadline a truck, the importance of moving a load must be weighed against the probability that a mechanical condition will result in a breakdown. The Federal Motor Carrier Safety Regulations, Part 393, say that anything mechanical that can either cause or prevent an accident is a safety item. All alleged safety problems should be addressed. The regulations also forbid dispatching a vehicle that is likely to break down. Regardless of whether any fine might be levied for that, dispatching a truck that is on the brink of breaking down is just not good business, nor is it safe.

    Answer B is incorrect. Items such as a malfunctioning cruise control or an engine brake are not deadline items and would not warrant deadlining the truck.

    Answer C is incorrect. Only Technician A is correct.

    Answer D is incorrect. Technician A is correct.

40. What should a technician do to thoroughly inspect a transmission mount?

    A. Remove the mount and tension the two mounting plates.

    B. Remove the mount and inspect for signs of damage.

    C. Visually inspect the mount while still in the vehicle.

    D. Replace the mount if it is suspect.

TASK D.2.8

    Answer A is incorrect. Removal of the mount should only be done if it is determined that the mount is defective by using proper troubleshooting techniques.

    Answer B is incorrect. Removal of the mount is not necessary to check for defectiveness.

    Answer C is correct. Visually inspect the mount while still in the vehicle. The most effective way of diagnosing a vehicle with defective engine mounts is to observe the engine mounts while torquing the engine with the brakes applied.

    Answer D is incorrect. Replacement of the mount should only be done after proper troubleshooting procedures have been followed and the mount is found to be defective.

TASK D.3.13

41. Technician A says that steer axle tires must have at least $^2/_{32}$ inch remaining at the most worn spot. Technician B says that tires on the drive or trailer axle must have at least $^4/_{32}$ inch tread remaining. Who is correct?

   A. A only

   B. B only

   C. Both A and B

   D. Neither A nor B

   Answer A is incorrect. Steer axle tires must have at least $^4/_{32}$ inch remaining at the most worn spot.

   Answer B is incorrect. Tires on the drive or trailer axle must have at least $^2/_{32}$ inch tread remaining.

   Answer C is incorrect. Neither Technician is correct.

   Answer D is correct. Neither Technician is correct. Measure the tread depth at the inboard, center, and outboard tread grooves, and at several places around the circumference of the tire.

42. Technician A says that EGR is an emissions device. Technician B says that an inoperative PCV can cause emission problems. Who is correct?

   A. A only

   B. B only

   C. Both A and B

   D. Neither A nor B

TASKS A.11, A.17

   Answer A is incorrect. Technician B is also correct.

   Answer B is incorrect. Technician A is also correct.

   Answer C is correct. Both Technicians are correct. Exhaust gas recirculation (EGR) is a method of reducing oxides of nitrogen by circulating a small amount of exhaust gases back through the intake manifold to dilute the intake charge and lower the combustion temperature. An inoperative positive crankcase ventilation (PCV) can cause erratic engine operation as well as higher than normal gasoline exhaust emissions.

   Answer D is incorrect. Both Technicians are correct.

TASK D.4.3

43. In what position should the height control valve lever be in when a truck with an air spring suspension system is at its normal height and the valve is in its neutral position?

   A. About 45 degrees upward from level

   B. Parallel to the ground

   C. About 45 degrees downward from level

   D. Approximately vertical

   Answer A is incorrect. 45 degrees upward from level would increase the inflation of the air springs.

   Answer B is correct. When the valve is in the neutral position, the lever should be level.

   Answer C is incorrect. 45 degrees downward from level would decrease the inflation of the air springs.

   Answer D is incorrect. This position is not possible.

44. What type of lubricant should be used to grease the fifth wheel?

   A. 80/90 gear lube
   B. Water-resistant lithium-based grease
   C. Any type of bearing grease
   D. 85/140 axle lube

TASK D.4.11

Answer A is incorrect. 80W-90 gear lube is typically used in drive axles.

Answer B is correct. The top surface of the fifth wheel must be lubricated with a water-resistant lithium-based grease. The consistency of grease, a measure of its relative hardness, is commonly expressed in terms of ASTM penetration or NLGI (National Lubricating Grease Institute) consistency number. Lubrication of the fifth-wheel top plate may be done by direct application or through the grease fittings on the underside of the top plate.

Answer C is incorrect. Bearing grease is not used to lubricate the fifth wheel.

Answer D is incorrect. 85/140 axle lube is not used to lubricate the fifth wheel.

45. Technician A says that saturated brake linings should be replaced. Technician B says that the friction rating of linings is coded by letter codes. Who is correct?

   A. A only
   B. B only
   C. Both A and B
   D. Neither A nor B

TASK D.1.15

Answer A is incorrect. Technician B is also correct.

Answer B is incorrect. Technician A is also correct.

Answer C is correct. Both Technicians are correct. Brake linings saturated with oil, grease, or brake fluid should be replaced. The friction rating of linings is coded by letter codes.

Answer D is incorrect. Both Technicians are correct.

46. What system could use a sliding gate valve?

   A. Cooling system
   B. Exhaust brake system
   C. A/C system
   D. Electrical system

TASK E.9

Answer A is incorrect. The cooling system does not use this type of valve.

Answer B is correct. The exhaust system uses a sliding gate valve in the exhaust brake. The sliding gate exhaust brake uses a pneumatically activated gate actuated by chassis system pressure. The air supply to close the gate is controlled by an electronically switched pilot valve. An aperture in the sliding gate permits minimal flow through the brake gate during engine braking. The butterfly valve version operates similarly.

Answer C is incorrect. The A/C system does not use this type of valve.

Answer D is incorrect. The electrical system does not use this type of valve.

TASKS D.3.17, D.3.15

47. Technician A says that spoke wheels generally experience greater alignment and balance problems than disc wheels. Technician B says that cracked wheels can be caused by vehicle overloading. Who is correct?

    A. A only
    B. B only
    C. Both A and B
    D. Neither A nor B

    Answer A is incorrect. Technician B is also correct.

    Answer B is incorrect. Technician A is also correct.

    Answer C is correct. Both Technicians are correct. Spoke wheels generally experience greater alignment and balance problems than disc wheels, but with proper installation and torquing, it is possible that spoke wheels can run virtually trouble-free. Cracked wheels can be caused by vehicle overloading.

    Answer D is incorrect. Both Technicians are correct.

TASK C.10

48. Technician A says that an improper ground could be the source of dim lights. Technician B says that continuity in an electrical circuit can be tested with a digital voltmeter/ohmmeter? Who is correct?

    A. A only
    B. B only
    C. Both A and B
    D. Neither A nor B

    Answer A is incorrect. Technician B is also correct.

    Answer B is incorrect. Technician A is also correct.

    Answer C is correct. Both Technicians are correct. A defective and/or improper ground connection are the most common sources of dim lights. Continuity in an electrical circuit can be tested with a digital voltmeter/ohmmeter.

    Answer D is incorrect. Both Technicians are correct.

2013 © Delmar, Cengage Learning

TASK A.4

49. What is shown in the illustration above?

    A. Cooling system tester
    B. Belt tension gauge
    C. Vacuum gauge
    D. Dial indicator

    Answer A is incorrect. A cooling system tester is hooked up to the inlet spout of the radiator.

    Answer B is correct. A belt tension gauge is shown in the illustration.

    Answer C is incorrect. A vacuum gauge uses a small vacuum hose for hook-up.

    Answer D is incorrect. A dial indicator does not have a handle but does have a small pointer.

50. Technician A says that each time the air compressor unloader cycles, it purges water and oil residue collected in the air dryer. Technician B says that air dryers discharge air and moisture into a holding tank. Who is correct?

TASK D.1.3

   A.   A only

   B.   B only

   C.   Both A and B

   D.   Neither A nor B

Answer A is correct. Only Technician A is correct. Water and oil residue collected in the air dryer are purged each time the air compressor unloader cycles, triggering the purge valve to eject the moisture.

Answer B is incorrect. Each time the unloader cycles, there should be discharge of air and moisture from the bottom of the air drier to atmosphere.

Answer C is incorrect. Only Technician A is correct.

Answer D is incorrect. Technician A is correct.

PREPARATION EXAM 6—ANSWER KEY

1.	B	18.	D	35.	C
2.	D	19.	A	36.	D
3.	A	20.	B	37.	D
4.	A	21.	C	38.	A
5.	C	22.	A	39.	B
6.	B	23.	A	40.	C
7.	A	24.	B	41.	A
8.	A	25.	D	42.	C
9.	D	26.	D	43.	B
10.	D	27.	C	44.	B
11.	D	28.	B	45.	C
12.	B	29.	C	46.	C
13.	B	30.	B	47.	D
14.	A	31.	A	48.	A
15.	A	32.	B	49.	D
16.	B	33.	A	50.	B
17.	C	34.	D		

PREPARATION EXAM 6—EXPLANATIONS

TASK C.3

1. Technician A says that a battery with a stabilized open circuit voltage of 12.2 V is about 25 percent charged. Technician B says that a battery with a stabilized open circuit voltage of 12.4 V is 75 percent charged. Who is correct?

A. A only

B. B only

C. Both A and B

D. Neither A nor B

Answer A is incorrect. A stabilized open circuit voltage of 12.2 V is 25 percent charged.

Answer B is correct. Only Technician B is correct. A stabilized open circuit voltage of 12.4 V is 75 percent charged.

Answer C is incorrect. Only Technician B is correct.

Answer D is incorrect. Technician B is correct.

2. When a self-adjusting clutch is found to be out of adjustment, check for all of the following EXCEPT:

   A. Correct actuator arm setting.

   B. Bent adjuster arm.

   C. Frozen adjusting ring.

   D. Clutch shaft runout.

TASKS D.2.2, D.2.3, D.2.4

Answer A is incorrect. Checking the placement of the actuator arm is part of a valid inspection. A self-adjusting clutch that is out of adjustment could be caused by an actuator arm incorrectly inserted into the release bearing sleeve retainer. Visually check to see that the actuator arm is correctly inserted. If the assembly is properly installed, the spring will move back and forth as the pedal is full stroked. The clutch will not compensate for wear if the actuator arm is not inserted into the release sleeve retainer, or if the release bearing travel is less than ½ inch.

Answer B is incorrect. Checking for a bent adjuster arm is a valid inspection. A self-adjusting clutch found out of adjustment could be caused by a bent adjuster arm.

Answer C is incorrect. Checking for a frozen adjusting ring is a valid inspection. The adjusting ring should be checked for damage or a frozen condition.

Answer D is correct. Clutch shaft runout is not checked on a self-adjusting clutch that is out of adjustment.

3. Technician A says moisture and corrosion on battery tops cause faster battery self-discharge. Technician B says excessive corrosion on the battery box and cover indicates undercharging. Who is correct?

   A. A only

   B. B only

   C. Both A and B

   D. Neither A nor B

TASK C.2

Answer A is correct. Only Technician A is correct. Moisture and corrosion on battery tops cause faster battery self-discharge.

Answer B is incorrect. Excessive corrosion on the battery box and cover indicates overcharging.

Answer C is incorrect. Only Technician A is correct.

Answer D is incorrect. Technician A is correct.

4. When testing the operation of the trailer control valve, a technician should do which of the following?

   A. Move the handle to the fully applied position and record the air pressure.

   B. Drain the air system to 0 psi, and then allow pressure to build up again.

   C. Record the gauge reading of the air pressure first, and then drain the system.

   D. Listen for air leakage around the handle as the handle is moved to the applied position.

TASK D.1.11

Answer A is correct. You do move the handle to the fully applied position and record the air pressure.

Answer B is incorrect. The air system does not have to be drained to test the trailer control valve.

Answer C is incorrect. As long as the trailer brakes are fully charged, the initial gauge reading is unnecessary.

Answer D is incorrect. Air leakage around the control handle should not occur.

TASK B.22

5. In winterizing a vehicle, Technician A will use a Teflon lubricant on door latches and dry graphite on the door locks. Technician B mounts alcohol evaporator downstream from the air dryer. Who is correct?

A. A only
B. B only
C. Both A and B
D. Neither A nor B

Answer A is incorrect. Technician B is also correct.

Answer B is incorrect. Technician A is also correct.

Answer C is correct. Both Technicians are correct. You use Teflon lubricant on the door latches and dry graphite on the door locks. You mount the alcohol evaporator downstream from the air dryer.

Answer D is incorrect. Both Technicians are correct.

TASK A.28

6. Oil that has a milky, clouded appearance is probably contaminated with which of the following?

A. Fuel
B. Engine coolant
C. Dust
D. Air

Answer A is incorrect. Fuel makes oil appear thinner and blacker in color.

Answer B is correct. Coolant in the engine lubricant gives it a milky, cloudy appearance.

Answer C is incorrect. Dust or dirt gives the oil a dirty and gritty appearance.

Answer D is incorrect. Air in the oil does not cause a milky or clouded appearance.

TASKS D.3.14,
D.3.15, D.3.16

7. Technician A says that wheel lug nuts should always be torqued. Technician B says that air pressure typically doesn't need to be rechecked once the tires are inflated to the proper air pressure. Who is correct?

A. A only
B. B only
C. Both A and B
D. Neither A nor B

Answer A is correct. Only Technician A is correct. Wheel nuts and lugs must be correctly torqued, and at recommended intervals, re-torquing is necessary.

Answer B is incorrect. Periodic checking of air pressure is the most effective means of prolonging tire wear and increases fuel mileage.

Answer C is incorrect. Only Technician A is correct.

Answer D is incorrect. Technician A is correct.

8. All of the following could cause sluggish service brake release EXCEPT:

A. Larger than specified brake chamber diameter.
B. Weak chamber return spring.
C. Broken or weak brake return springs.
D. Obstructed brake chamber air passage.

Answer A is correct. A brake chamber with a larger than specified diameter increases braking force, but it does not affect brake release time.

Answer B is incorrect. A weak chamber return spring causes sluggish brake release.

Answer C is incorrect. Broken or weak brake return springs cause sluggish brake release.

Answer D is incorrect. An obstruction in a brake chamber air passage causes sluggish brake release.

TASK D.1.14

9. Technician A says that pour point refers to the temperature at which the normal paraffins in the fuel become less soluble and begin to precipitate as wax crystals. Technician B says that the cloud point is generally 5 °F to 25 °F (3 °C to 15 °C) below the pour point. Who is correct?

TASK A.10

   A. A only

   B. B only

   C. Both A and B

   D. Neither A nor B

Answer A is incorrect. Pour point of a fuel is generally considered to be 5 °F above the temperature at which the fuel will no longer flow.

Answer B is incorrect. The pour point is generally 5 °F to 25 °F (3 °C to 15 °C) below cloud point.

Answer C is incorrect. Neither Technician is correct.

Answer D is correct. Neither Technician is correct. The pour point is the lowest temperature at which diesel fuel will flow. This temperature is lower than the cloud point and is easier to modify or lower using additives. The cloud point is the temperature at which a haze of wax crystals is formed. Cloud point is more significant than pour point when determining whether a fuel can be pumped through filters in cold weather.

10. A transmission temperature gauge does not operate. Technician A says that the sensor should be replaced because he sees the same problem recur frequently. Technician B reads a value of 1 volt at the electrical connector for the sensor and replaces the sensor because the wiring must be intact. Who is correct?

TASK D.2.6

   A. A only

   B. B only

   C. Both A and B

   D. Neither A nor B

Answer A is incorrect. It is not good practice to replace parts before finding the fault.

Answer B is incorrect. Most sensors need at least 5 volts to operate properly. Some instrument panel gauges require protection against heavy voltage fluctuations that could damage the gauges or cause them to give incorrect readings. A voltage limiter provides this protection by limiting voltage to the gauges to a preset value. The limiter contains a heating coil, a bimetal arm, and a set of contacts. When the ignition is in the on or accessory position, the heating coil heats the bimetal arm, causing it to bend and open the contacts. This action cuts the voltage from both the heating coil and the circuit. When the arm cools down to the point where the contacts close, the cycle repeats itself. Rapid opening and closing of the contacts produces a pulsating voltage at the output terminal that averages 5 volts. Voltage limiting may also be performed electronically.

Answer C is incorrect. Neither Technician is correct.

Answer D is correct. Neither Technician is correct. Always perform appropriate troubleshooting procedures before deeming a part in need of replacement.

TASK B.31

11. Filters installed in the HVAC air delivery system are:

A. Always made of fiberglass mesh to resist corrosion.
B. Designed to remove moisture from cab and sleeper air.
C. Not individually replaceable.
D. Designed to remove dust and dirt from cab and sleeper air.

Answer A is incorrect. Some evaporator filters are made of paper or a metallic mesh.

Answer B is incorrect. Evaporator filters are not designed to remove moisture from the air.

Answer C is incorrect. Evaporator filters are replaceable.

Answer D is correct. They are designed to remove dust and dirt from cab and sleeper air.

TASKS B.23, B.26

12. Technician A says air seat spring suspensions eliminate the need for cab shock absorbers. Technician B says you can adjust the leveling valve to maintain the proper cab height. Who is correct?

A. A only
B. B only
C. Both A and B
D. Neither A nor B

Answer A is incorrect. Air seat spring suspensions do not eliminate the need for cab shock absorbers.

Answer B is correct. Only Technician B is correct. You can adjust the leveling valve to maintain the proper cab height.

Answer C is incorrect. Only Technician B is correct.

Answer D is incorrect. Technician B is correct.

TASK C.8

13. When testing the maximum output of an alternator, what should the engine be run at?

A. 1,000 rpm
B. 1,500–2,000 rpm
C. Idle speed
D. 2,000 alternator rpm

Answer A is incorrect. 1,000 rpm would usually be too slow for maximum alternator output.

Answer B is correct. Alternator output is usually tested at 1,500–2,000 rpm.

Answer C is incorrect. The alternator will not produce maximum output at idle speed.

Answer D is incorrect. An alternator produces maximum output at 1,500–2,000 rpm.

TASKS B.10, B.11

14. The seat belts in the cab and sleeper are being inspected. Technician A says with the seat belt on and the truck moving, activate the parking brake to stop the truck. Technician B says it is not necessary to check the seat belts or restraints in the sleeper, only the cab. Who is correct?

A. A only
B. B only
C. Both A and B
D. Neither A nor B

Answer A is correct. Only Technician A is correct. Activating the parking brake to brake the truck very quickly at low speeds will determine if the seat belt inertia reel locks.

Answer B is incorrect. All seat belts and restraints of any type must be checked on a PMI.

Answer C is incorrect. Only Technician A is correct.

Answer D is incorrect. Technician A is correct.

15. What is the minimum performance (psi/sec) a compressor should have?

    A. 85–100 psi in 25 seconds or less

    B. 60–80 psi in 20 seconds or less

    C. 90–110 psi in 30 seconds or less

    D. 50–110 psi in 25 seconds or less

TASK D.1.10

Answer A is correct. 85–100 psi in 25 seconds or less is the anticipated specification.

Answer B is incorrect. 60–80 psi in 20 seconds is not the generally accepted standard for compressor performance.

Answer C is incorrect. 90–110 in 30 seconds is not the generally accepted standard for compressor performance.

Answer D is incorrect. 50–110 psi in 25 seconds is not the generally accepted standard for compressor performance.

16. Technician A says that an oil sample should be taken once the oil is drained from the engine. Technician B says that an oil sample can test for contamination, metal, and dirt content. Who is correct?

    A. A only

    B. B only

    C. Both A and B

    D. Neither A nor B

TASK A.28

Answer A is incorrect. An oil sample should be taken during an oil change at the midpoint of the drainage.

Answer B is correct. Only Technician B is correct. An oil sample can test for contamination, metal, and dirt content.

Answer C is incorrect. Only Technician B is correct.

Answer D is incorrect. Technician B is correct.

17. A check of the general condition of the alternator is being performed. Technician A says to check for out-of-round mounting holes. Technician B checks the belts before making performance tests. Who is correct?

    A. A only

    B. B only

    C. Both A and B

    D. Neither A nor B

TASK C.7

Answer A is incorrect. Technician B is also correct.

Answer B is incorrect. Technician A is also correct.

Answer C is correct. Both Technicians are correct. The alternator should be checked for out-of-round mounting holes. Alternator belts should be checked for proper tightness before making performance tests.

Answer D is incorrect. Both Technicians are correct.

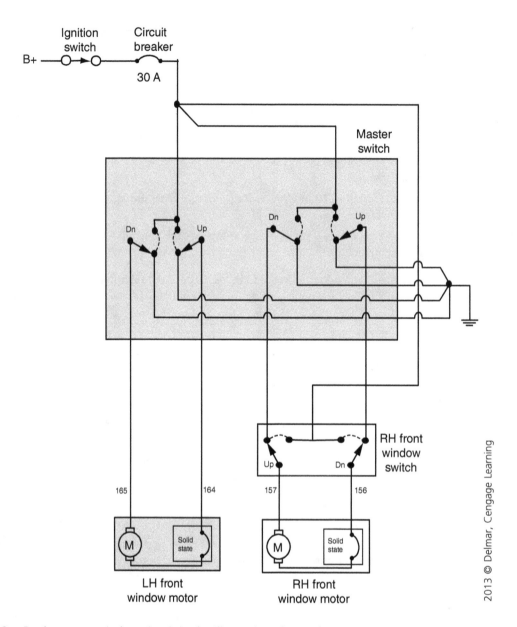

2013 © Delmar, Cengage Learning

**TASKS C.10,
B.17, B.19**

18. In the power window circuit in the illustration above, there is an open circuit in wire 164
     between the master switch and the LH power window motor. The result of this defect is:

     A. The LH window will go down, but it will not go back up.

     B. The RH window will go down, but it will not go back up.

     C. The LH window will stall part way down.

     D. The LH window is inoperative.

     Answer A is incorrect. The LH window will be inoperative.

     Answer B is incorrect. The LH window will be inoperative.

     Answer C is incorrect. The LH window will be inoperative.

     Answer D is correct. The LH window will be inoperative.

19. During brake inspection, a technician tests the air supply system and finds that the buildup is slow. Technician A says the compressor outlet hose may be restricted. Technician B says there may be a leak in one of the brake chambers. Who is correct?

    A. A only

    B. B only

    C. Both A and B

    D. Neither A nor B

TASK D.1.10

Answer A is correct. Only Technician A is correct. A restricted compressor outlet hose causes slow compressor buildup time.

Answer B is incorrect. A leaking brake chamber diaphragm does not cause slow compressor buildup time, because there is no air pressure supplied to the brake chamber diaphragms until the brakes are applied.

Answer C is incorrect. Only Technician A is correct.

Answer D is incorrect. Technician A is correct.

20. A tractor with air-spring suspension sits low when coupled to a trailer and sits high without the trailer. Technician A says a leaking air spring could be the problem. Technician B says an improperly adjusted height control valve could be the problem. Who is correct?

    A. A only

    B. B only

    C. Both A and B

    D. Neither A nor B

TASK D.4.3

Answer A is incorrect. A leaking air spring will cause the trailer to sit low, not high.

Answer B is correct. Only Technician B is correct. An improperly adjusted height control valve could be the problem.

Answer C is incorrect. Only Technician B is correct.

Answer D is incorrect. Technician B is correct.

21. The normal system pressure for a cooling system is in the range of:

    A. 8 psi.

    B. 12 psi.

    C. 15 psi.

    D. 5 psi.

TASK A.20

Answer A is incorrect. The normal system pressure for a cooling system is in the range of 15 psi.

Answer B is incorrect. The normal system pressure for a cooling system is in the range of 15 psi.

Answer C is correct. The normal system pressure for a cooling system is in the range of 15 psi.

Answer D is incorrect. The normal system pressure for a cooling system is in the range of 15 psi.

TASK A.13

22. Technician A says that a minor exhaust leak under the cab is cause for deadline of the vehicle under Commercial Vehicle Safety Alliance (CVSA) criteria. Technician B says that if the exhaust contact with air lines or any wiring is permissible as long as damage is not present. Who is correct?

   A. A only

   B. B only

   C. Both A and B

   D. Neither A nor B

Answer A is correct. Only Technician A is correct. Any exhaust system that is leaking in front of or below the driver/sleeper compartment and the floor pan has potential for cab entry of exhaust fumes and the CVSA out-of-service criteria would deadline a vehicle for this.

Answer B is incorrect. The exhaust routing must be checked and assembled in a way that does not allow contact of the exhaust system with any wiring, air lines, or fuel lines, and the CVSA out-of-service criteria would deadline a vehicle for any of these faults.

Answer C is incorrect. Only Technician A is correct.

Answer D is incorrect. Technician A is correct.

TASKS E.1,
B.25

23. A truck's clutch is being measured for its free travel. Technician A says the measurement should be between 1 and 2 inches. Technician B says that on a hydraulic clutch, the measurement should be 3 to 4 inches. Who is correct?

   A. A only

   B. B only

   C. Both A and B

   D. Neither A nor B

Answer A is correct. Only Technician A is correct. The measurement should be between 1 and 2 inches.

Answer B is incorrect. The manufacturer's recommendations should always be checked on hydraulic clutches.

Answer C is incorrect. Only Technician A is correct.

Answer D is incorrect. Technician A is correct.

24. A driveshaft is being inspected by a technician during a PMI. Technician A says the transmission output shaft and rear-axle input shaft do not need to be checked. Technician B always checks the U-joint trunnions for cracks, the tubes for dents, and the center bearing for damage. Who is correct?

TASK D.2.10

   A. A only

   B. B only

   C. Both A and B

   D. Neither A nor B

Answer A is incorrect. When inspecting the driveline, the transmission output shaft and rear-axle input shaft should be checked because they can affect the shaft's harmonics. End yokes should be checked for retaining nut looseness, radial looseness, fractures, runout, and vertical alignment, which could cause vibration, premature wear of Powertrain components, and failure of U-joints.

Answer B is correct. Only Technician B is correct. A technician should always check the trunnions, tubes, and center support bearing for damage. Common U-joint damage includes cracking of trunnions or spider, galling, spalling, pitting, and brinelling. Check the slip spline for excessive radial movement. Check the shaft for damage, bent tubing, missing balance weights, or foreign material on the shaft. The center bearing should be checked for leaking lubricant, damaged seals, and proper mounting.

Answer C is incorrect. Only Technician B is correct.

Answer D is incorrect. Technician B is correct.

25. Technician A says that typical air filter restriction specifications are 10 inches of water vacuum for boosted engines. Technician B says that typical air filter restrictions are 30 inches of water vacuum for naturally aspirated engines. Who is correct?

TASK A.16

   A. A only

   B. B only

   C. Both A and B

   D. Neither A nor B

Answer A is incorrect. Typical air filter restriction specifications are 25 inches of water vacuum for boosted engines.

Technician B is incorrect. Typical air filter restrictions are 15 inches of water vacuum for naturally aspirated engines.

Answer C is incorrect. Neither Technician is correct.

Answer D is correct. Neither Technician is correct. An air filter restriction specification of 10 inches of water vacuum for a boosted engine would be well below the actual specification of 25 inches. An air filter restriction specification of 30 inches would be well above the recommended specification of 15 inches of water vacuum for a naturally aspirated engine.

TASK B.29

26. Which leak detection method is the least accurate?

   A.   Electronic leak detectors

   B.   Ultraviolet (UV) leak detectors

   C.   Fluid leak detectors

   D.   Soapsuds solution

Answer A is incorrect. Electronic leak detection is the most sensitive of the leak detection methods.

Answer B is incorrect. UV detectors are capable of detecting small leaks and can be used for hard to find or pinpoint leaks, especially cold or hot leaks. It is a very accurate leak detection method.

Answer C is incorrect. Commercial liquid leak detectors perform similar to the soapsuds solution method, but can locate smaller leaks.

Answer D is correct. The soapsuds solution method is the least accurate of the methods listed. This method is used when no other leak detection equipment is available.

TASK C.2

27. Technician A says that a voltage drop test can be used to check the condition of battery cables. Technician B says that resistance generates heat, which further increases resistance. Who is correct?

   A.   A only

   B.   B only

   C.   Both A and B

   D.   Neither A nor B

Answer A is incorrect. Technician B is also correct.

Answer B is incorrect. Technician A is also correct.

Answer C is correct. Both Technicians are correct. A voltage drop test can be used to check the condition of battery cables. Resistance generates heat, which further increases resistance.

Answer D is incorrect. Both Technicians are correct.

TASK A.12

28. An exhaust system on a cab-over is being inspected. Technician A says to inspect the system while it is hot, so proper pipe to body/frame clearances may be measured. Technician B uses a prybar to check the looseness of the piping and the strength of the brackets when performing an exhaust system PMI. Who is correct?

   A.   A only

   B.   B only

   C.   Both A and B

   D.   Neither A nor B

Answer A is incorrect. It is not necessary to check an exhaust system while it is hot.

Answer B is correct. Only Technician B is correct. A prybar can be used to check the looseness of the piping and the strength of the brackets when performing an exhaust system PMI.

Answer C is incorrect. Only Technician B is correct.

Answer D is incorrect. Technician B is correct.

29. A truck with an air brake system has reduced braking force on one rear wheel. Which of the following is the LEAST LIKELY cause?

TASK D.1.15

   A.   Restricted airflow or low air pressure to the service brake chamber
   B.   Improper adjustment of slack adjuster and brake chamber pushrod
   C.   Grease on shoe linings/faces
   D.   A ruptured brake chamber diaphragm

   Answer A is incorrect. A condition causing a lowering of air pressure will subsequently reduce service brake application force.

   Answer B is incorrect. An incorrect or lack of adjustment to a slack adjuster will cause a reduction in overall brake application forces.

   Answer C is correct. Grease on shoe linings/faces causes grabbing. It is the least likely cause of insufficient braking.

   Answer D is incorrect. A ruptured brake chamber diaphragm will severely decrease an air brake system's service application force.

30. Technician A says that alternator output should be within 20 percent of the specified rating. Technician B says that voltage output of the alternator is measured with the engine running. Who is correct?

TASKS C.8, C.9

   A.   A only
   B.   B only
   C.   Both A and B
   D.   Neither A nor B

   Answer A is incorrect. Alternator output should be within 10 percent of the specified rating.

   Answer B is correct. Only Technician B is correct. Alternator voltage output is measured with the engine running.

   Answer C is incorrect. Only Technician B is correct.

   Answer D is incorrect. Technician B is correct.

31. Technician A says that the best method for checking a steering system is with the wheels on the shop floor and with the engine running. Technician B says that manual steering wheel play should not exceed 4 inches. Who is correct?

TASK D.3.1

   A.   A only
   B.   B only
   C.   Both A and B
   D.   Neither A nor B

   Answer A is correct. Only Technician A is correct. The best method for checking a steering system is with the wheels on the floor and the engine running.

   Answer B is incorrect. The manual steering wheel play should not exceed 2 inches.

   Answer C is incorrect. Only Technician A is correct.

   Answer D is incorrect. Technician A is correct.

TASK D.1.3

32. All of the statements about air brake supply systems are true EXCEPT:

    A. The supply reservoir tends to collect more moisture than the front or rear axle reservoirs.

    B. Water must be drained daily from the reservoirs if an air dryer is used in the air brake system.

    C. A small amount of oil in the supply reservoir is a normal condition.

    D. Some air dryers and automatic moisture ejectors contain electric heaters to prevent moisture from freezing in these components.

    Answer A is incorrect. The supply reservoir tends to collect more moisture than the front or rear axle reservoirs.

    Answer B is correct. Water does not need to be drained daily from the reservoirs if an air dryer is used in the air brake system.

    Answer C is incorrect. A small amount of oil in the supply reservoir is a normal condition.

    Answer D is incorrect. Some air dryers and automatic moisture ejectors contain electric heaters to prevent moisture from freezing in these components.

TASK C.3, C.4

33. Technician A says that a load test should be performed after the state of charge test. Technician B says that the state of charge of a fully charged battery will be 12.4 volts. Who is correct?

    A. A only

    B. B only

    C. Both A and B

    D. Neither A nor B

    Answer A is correct. Only Technician A is correct. The state of charge test is performed before the battery load test.

    Answer B is incorrect. The state of charge of a fully charged battery will be 12.6 volts or higher.

    Answer C is incorrect. Only Technician A is correct.

    Answer D is incorrect. Technician A is correct.

2013 © Delmar, Cengage Learning

34. What is the component marked "B" in the illustration above?

A. S-cam

B. Power spring

C. Check valve

D. Pressure plate

TASK D.1.13

Answer A is incorrect. Component B is the pressure plate.

Answer B is incorrect. Component B is the pressure plate.

Answer C is incorrect. Component B is the pressure plate.

Answer D is correct. Component B is the pressure plate.

35. Technician A says that when disconnecting a battery, the ground cable should always be removed first. Technician B says that when reconnecting a battery, the positive cable should always be connected first. Who is correct?

A. A only

B. B only

C. Both A and B

D. Neither A nor B

TASK C.2

Answer A is incorrect. Technician B is also correct.

Answer B is incorrect. Technician A is also correct.

Answer C is correct. Both Technicians are correct. When disconnecting a battery, the ground cable should always be removed first. When reconnecting a battery, the positive cable should always be connected first.

Answer D is incorrect. Both Technicians are correct.

TASK E.5

36. Technician A says that an inspection of the road speed governor is required during the pre-trip inspection. Technician B says that the road speed governor is not programmable. Who is correct?

A.   A only

B.   B only

C.   Both A and B

D.   Neither A nor B

Answer A is incorrect. The operation of a road speed governor, or limiter, cannot be checked on a pre-trip inspection. Checking the road speed limiter is part of a vehicle road test. Connect the appropriate engine scan tool, OEM special tool, or laptop computer to the truck's diagnostic connector. During the road test, activate the tool's parameter specification readout and record the road speed. Reprogram if required.

Answer B is incorrect. The purpose of a road speed governor, or limiter, is to control the top speed of a vehicle. Limiting the speed of a vehicle increases fuel efficiency and promotes safety. The speed limiter is a programmable module, which may be reprogrammed by the driver.

Answer C is incorrect. Neither Technician is correct.

Answer D is correct. Neither Technician is correct.

2013 © Delmar, Cengage Learning

TASKS D.1.17, D.2.4

37. The slack adjuster in the illustration above should be adjusted so the brake chamber pushrod is:

A.   At an 85-degree angle in relation to the slack adjuster arm when fully applied.

B.   Parallel to the slack adjuster arm when the brakes are fully applied.

C.   At a 95-degree angle in relation to the slack adjuster arm when fully applied.

D.   At a 90-degree angle in relation to the slack adjuster arm when fully applied.

Answer A is incorrect. The overall braking force is decreased if the angle is less than 90 degrees between the brake chamber pushrod and the slack adjuster arm.

Answer B is incorrect. A brake chamber pushrod cannot be parallel to the slack adjuster to be functional.

Answer C is incorrect. Overall braking force is decreased if there is more than a 90-degree angle between the bake chamber pushrod and the slack adjuster arm.

Answer D is correct. The braking force is at the maximum if there is a 90-degree angle between the brake chamber pushrod and the slack adjuster arm.

38. While inspecting a manual transmission for leaks, the technician notices that some gaskets appear to have blown out of their mating surfaces. What should the technician check first?

    A. The transmission breather

    B. The shift housing cover

    C. The shift tower

    D. The rear seal

Answer A is correct. When the transmission breather is plugged, internal pressure can build inside the transmission and push oil out seals and gaskets. Always check the transmission breather filters when repairing any transmission leaks. Under normal operating conditions, the transmission can build up pressure inside the case if the vents or breathers are not functioning correctly. This pressure can cause fluid to leak past seals that do not need replacement or seals that are otherwise in good working order.

Answer B is incorrect. The breather could be plugged. The most common cause of gaskets appearing to be blown out of the shift housing cover would be a plugged breather.

Answer C is incorrect. The breather could be plugged. The most common cause of gaskets appearing to be blown out of the shift tower would be a plugged breather.

Answer D is incorrect. The breather could be plugged. When a breather becomes plugged, fluid is often forced past seals in the transmission.

39. A tire has a cut in the sidewall that shows exposed tire cords. Technician A says that the cut should be patched. Technician B says that the tire must be replaced. Who is correct?

    A. A only

    B. B only

    C. Both A and B

    D. Neither A nor B

Answer A is incorrect. A tire patch in not an acceptable repair method for a tire with a cut that shows exposed tire cords and the tire must be replaced for safety reason.

Answer B is correct. Only Technician B is correct. The tire must be replaced for safety reasons.

Answer C is incorrect. Only Technician B is correct.

Answer D is incorrect. Technician B is correct.

40. Technician A says that a loose steering gear mount could be binding the steering column and cause the power steering to become increasingly harder to turn. Technician B says that power steering wheel play should not exceed 2 ½ inches. Who is correct?

    A. A only

    B. B only

    C. Both A and B

    D. Neither A nor B

Answer A is incorrect. Technician B is also correct.

Answer B is incorrect. Technician A is also correct.

Answer C is correct. Both Technicians are correct. A loose steering gear mount could be binding the steering column and cause the power steering to become increasingly harder to turn. Power steering wheel play should not exceed 2 ½ inches.

Answer D is incorrect. Both Technicians are correct.

TASK D.4.12

41. All of the following practices should be avoided when working with truck frames EXCEPT:

 A. Drilling holes in the frame web.
 B. Drilling holes in the frame flanges.
 C. Cutting holes in the frame with a torch.
 D. Notching the frame rails.

 Answer A is correct. The frame web may be drilled, but careful consideration should be given to the number, location, and sizes of additional frame bolt holes. Locate holes as close as possible to the horizontal center of the web, and do not place more than three bolt holes on the same vertical line in the web. Do not drill holes in heat-treated frames.

 Answer B is incorrect. Holes should never be drilled into the flange sections of the frame rails, as this weakens the frame.

 Answer C is incorrect. Cutting holes in the frame with a torch is not an accepted practice. Bolt holes should not be larger than the size of the bolt being installed.

 Answer D is incorrect. Notching the frame rails severely weakens the structure and could ultimately result in fracture and subsequent failure.

TASK A.27

42. Technician A says that a milky gray sludge indicates water in the oil. Technician B says that used oil and filters are considered hazardous waste. Who is correct?

 A. A only
 B. B only
 C. Both A and B
 D. Neither A nor B

 Answer A is incorrect. Technician B is also correct.

 Answer B is incorrect. Technician A is also correct.

 Answer C is correct. Both Technicians are correct. A milky gray sludge indicates water in the oil caused by a possible head or cylinder block leak. Used oil and filters are considered hazardous waste and must be disposed of in accordance with federal, state, and local regulations.

 Answer D is incorrect. Both Technicians are correct.

TASKS D.4.1, D.4.2, D.4.3

43. When testing a suspension system on a heavy-duty truck, a technician should do which of the following?

 A. Bounce the vehicle up and down to test the shocks.
 B. Visually inspect the suspension components for signs of fatigue, deterioration, and/or failure.
 C. Road test the vehicle to determine possible defects and/or misalignment(s).
 D. Check the PMI schedule for scheduled repairs.

 Answer A is incorrect. It is not possible to "bounce" a heavy truck up and down to test its shocks. Even if it bounced, it would not be a good indicator of shock wear.

 Answer B is correct. A technician should give the truck a thorough visual inspection of all suspension components for signs of fatigue, deterioration, and/or failure. On leaf spring suspensions, check spring U-bolts, spring pins and bushings, shackles, spring brackets, hangers, and spring assemblies. Springs should be inspected for cracks, sag, and surface wear. Multiple leaf spring assemblies should be replaced as an assembly. On air suspension systems, check air springs, mounts, hoses, valves, linkage, and fittings for leaks or damage.

 Answer C is incorrect. A road test will not always give accurate results of the condition of a suspension or even steering linkage.

 Answer D is incorrect. A check of a PMI schedule for repairs should only be made after a thorough inspection on the status of the vehicle.

44. Technician A says that the purpose of the radiator cap is to allow lower operating temperatures. Technician B says that an underperforming radiator cap may cause engine overheating and coolant loss. Who is correct?

TASK A.20

    A.   A only

    B.   B only

    C.   Both A and B

    D.   Neither A nor B

Answer A is incorrect. The radiator cap allows for higher operating temperatures.

Answer B is correct. Only Technician B is correct. An underperforming radiator cap may cause overheating and coolant loss.

Answer C is incorrect. Only Technician B is correct.

Answer D is incorrect. Technician B is correct.

45. A slack adjuster is being adjusted. Technician A says excessive wear limits between the clevis and the collar will cause the stroke to be too long. Technician B says improper brake adjustment may cause premature wheel lockup during a brake application. Who is correct?

TASKS D.1.14, D.1.17

    A.   A only

    B.   B only

    C.   Both A and B

    D.   Neither A nor B

Answer A is incorrect. Technician B is also correct.

Answer B is incorrect. Technician A is also correct.

Answer C is correct. Both Technicians are correct. Excessive wear limits between the clevis and the collar will cause the stroke to be too long. Improper brake adjustment may cause premature wheel lockup during brake application,

Answer D is incorrect. Both Technicians are correct.

46. A gauge cluster is being inspected for proper function and illumination. Technician A says that a malfunctioning oil pressure gauge is not a valid reason for tagging a vehicle out of service. Technician B says that a proper function check of the instrumentation panel is required under a PMI. Who is correct?

TASK E.2

    A.   A only

    B.   B only

    C.   Both A and B

    D.   Neither A nor B

Answer A is wrong. Technician B is also correct.

Answer B is wrong. Technician A is also correct.

Answer C is correct. Both Technicians are correct. A malfunctioning oil pressure gauge is not a valid reason for tagging a vehicle out of service. A proper function check of the instrumentation panel is required under a PMI.

Answer D is wrong. Both Technicians are correct.

TASK D.4.1

47. All of the following spring types can be found on a heavy-duty truck suspension EXCEPT:

   A.   Constant rate.
   B.   Progressive rate.
   C.   Auxiliary.
   D.   Coil.

Answer A is incorrect. Constant rate springs are leaf-type assemblies that have a constant rate of deflection.

Answer B is incorrect. Progressive rate springs are leaf-type spring assemblies with a variable deflection rate obtained by varying the effective length of the spring assembly.

Answer C is incorrect. Auxiliary springs are leaf-type spring assemblies usually mounted on top of the truck rear spring assemblies to increase load-carrying ability.

Answer D is correct. Coil springs are the exception because they are not used on heavy-duty truck suspensions.

Brakes applied

Brakes released

Brake shoe
rollers

S-cam

Brake shoes

2013 © Delmar, Cengage Learning

48. The brake linings are worn and there is excessive clearance between the linings and the brake
drum on a cam-type foundation brake (see illustration above). Technician A says this
condition may cause the cam to roll over during a brake application. Technician B says this
condition may cause loss of steering control. Who is correct?

TASK D.1.15

A. A only

B. B only

C. Both A and B

D. Neither A nor B

Answer A is correct. Only Technician A is correct. Excessive clearance between the linings
and the brake drum may cause the cam to roll over during a brake application.

Answer B is incorrect. This condition will not cause loss of steering control.

Answer C is incorrect. Only Technician A is correct.

Answer D is incorrect. Technician A is correct.

TASK E.7

49. Technician A says that black smoke is normal and not a cause for concern. Technician B says black smoke is a sign that the engine is getting the correct amount of fuel and air for optimum performance. Who is correct?

  A. A only
  B. B only
  C. Both A and B
  D. Neither A nor B

Answer A is incorrect. Black smoke is not a normal operating condition.

Answer B is incorrect. Black smoke is not a sign of optimum air/fuel ratio.

Answer C is incorrect. Neither Technician is correct.

Answer D is correct. Neither Technician is correct. The color and consistency of the exhaust can reveal combustion problems. When a diesel engine is running properly with a heavy load, exhaust emission should be clear. Exhaust smoke can be an indication of serious problems.

TASK A.3

50. Technician A says vibration damper absorbs camshaft harmonic distortions. Technician B says the vibration damper balances harmonic vibrations caused by torsional loads of the rotating crankshaft. Who is correct?

  A. A only
  B. B only
  C. Both A and B
  D. Neither A nor B

Answer A is incorrect. The camshaft does not use a vibration damper.

Answer B is correct. Only Technician B is correct. The vibration damper balances harmonic vibrations caused by torsional loads of the rotating crankshaft.

Answer C is incorrect. Only Technician B is correct.

Answer D is incorrect. Technician B is correct.

Appendices

PREPARATION EXAM ANSWER SHEET FORMS

ANSWER SHEET

1. _____	21. _____	41. _____
2. _____	22. _____	42. _____
3. _____	23. _____	43. _____
4. _____	24. _____	44. _____
5. _____	25. _____	45. _____
6. _____	26. _____	46. _____
7. _____	27. _____	47. _____
8. _____	28. _____	48. _____
9. _____	29. _____	49. _____
10. _____	30. _____	50. _____
11. _____	31. _____	
12. _____	32. _____	
13. _____	33. _____	
14. _____	34. _____	
15. _____	35. _____	
16. _____	36. _____	
17. _____	37. _____	
18. _____	38. _____	
19. _____	39. _____	
20. _____	40. _____	

ANSWER SHEET

1. _____		21. _____		41. _____	
2. _____		22. _____		42. _____	
3. _____		23. _____		43. _____	
4. _____		24. _____		44. _____	
5. _____		25. _____		45. _____	
6. _____		26. _____		46. _____	
7. _____		27. _____		47. _____	
8. _____		28. _____		48. _____	
9. _____		29. _____		49. _____	
10. _____		30. _____		50. _____	
11. _____		31. _____			
12. _____		32. _____			
13. _____		33. _____			
14. _____		34. _____			
15. _____		35. _____			
16. _____		36. _____			
17. _____		37. _____			
18. _____		38. _____			
19. _____		39. _____			
20. _____		40. _____			

ANSWER SHEET

1. _____	21. _____	41. _____
2. _____	22. _____	42. _____
3. _____	23. _____	43. _____
4. _____	24. _____	44. _____
5. _____	25. _____	45. _____
6. _____	26. _____	46. _____
7. _____	27. _____	47. _____
8. _____	28. _____	48. _____
9. _____	29. _____	49. _____
10. _____	30. _____	50. _____
11. _____	31. _____	
12. _____	32. _____	
13. _____	33. _____	
14. _____	34. _____	
15. _____	35. _____	
16. _____	36. _____	
17. _____	37. _____	
18. _____	38. _____	
19. _____	39. _____	
20. _____	40. _____	

ANSWER SHEET

1. _____	21. _____	41. _____
2. _____	22. _____	42. _____
3. _____	23. _____	43. _____
4. _____	24. _____	44. _____
5. _____	25. _____	45. _____
6. _____	26. _____	46. _____
7. _____	27. _____	47. _____
8. _____	28. _____	48. _____
9. _____	29. _____	49. _____
10. _____	30. _____	50. _____
11. _____	31. _____	
12. _____	32. _____	
13. _____	33. _____	
14. _____	34. _____	
15. _____	35. _____	
16. _____	36. _____	
17. _____	37. _____	
18. _____	38. _____	
19. _____	39. _____	
20. _____	40. _____	

ANSWER SHEET

1. _____	21. _____	41. _____
2. _____	22. _____	42. _____
3. _____	23. _____	43. _____
4. _____	24. _____	44. _____
5. _____	25. _____	45. _____
6. _____	26. _____	46. _____
7. _____	27. _____	47. _____
8. _____	28. _____	48. _____
9. _____	29. _____	49. _____
10. _____	30. _____	50. _____
11. _____	31. _____	
12. _____	32. _____	
13. _____	33. _____	
14. _____	34. _____	
15. _____	35. _____	
16. _____	36. _____	
17. _____	37. _____	
18. _____	38. _____	
19. _____	39. _____	
20. _____	40. _____	

ANSWER SHEET

1. _____	21. _____	41. _____
2. _____	22. _____	42. _____
3. _____	23. _____	43. _____
4. _____	24. _____	44. _____
5. _____	25. _____	45. _____
6. _____	26. _____	46. _____
7. _____	27. _____	47. _____
8. _____	28. _____	48. _____
9. _____	29. _____	49. _____
10. _____	30. _____	50. _____
11. _____	31. _____	
12. _____	32. _____	
13. _____	33. _____	
14. _____	34. _____	
15. _____	35. _____	
16. _____	36. _____	
17. _____	37. _____	
18. _____	38. _____	
19. _____	39. _____	
20. _____	40. _____	

Glossary

ABS Acronym for antilock brake system.

A/D Converter Abbreviation for analog-to-digital converter.

After-Cooler Can be used to describe a variety of different heat exchangers used on engines, transmissions, and air brake systems. An after-cooler on an air brake system is used to condense water from moisture-saturated air leaving the compressor by cooling it.

Air Brakes Pneumatically actuated brakes used on most heavy-duty trucks; uses air pressure to brake the vehicle by means of diaphragms, levers, S-cams, wedges, and rotor assemblies.

Air Brake System A system utilizing compressed air to activate the brakes.

Air Compressor An engine-driven mechanism for supplying air to the truck brake system.

Air Dryer A device usually located downstream from the compressor designed to remove moisture from the air system.

Air Filter/Regulator Assembly A device that minimizes the possibility of moisture-laden air or impurities entering a system, which also regulates the air pressure in the system.

Air-Over-Hydraulic Brakes A brake system utilizing a hydraulic system assisted by an air pressure system.

Air Shifting The process that uses air pressure to engage different range of combinations in the transmission's auxiliary section without a mechanical linkage to the driver.

Air-Spring Suspension A single or multi-axle suspension relying on airbags for springs and weight distribution of axles.

Ambient Temperature Temperature of the surrounding or prevailing air. Normally, it is considered to be the temperature in the service area where testing is taking place.

Ampere (amp) The unit for measuring electrical current.

Analog Volt/Ohmmeter (AVOM) A test meter used for checking voltage and resistance. Not usually suited for use on digital circuits.

Anti Rattle Springs Springs that reduce wear between the intermediate plate and the drive pin and help improve clutch release.

Antilock Brake System (ABS) An electronically managed hydraulic or pneumatic air brake system that senses wheel speeds and manages application pressures to prevent wheel lockup.

Antilock Relay Valve (ARV) In an antilock brake system, the device that usually replaces the standard relay valve used to control the rear-axle service brakes and performs the standard relay function during tractor/trailer operation.

Application Valve A foot-operated brake valve that controls air pressure to the service chambers.

Armature The rotating component of a (1) starter or other motor, (2) generator, and (3) compressor clutch.

ASE Acronym for Automotive Service Excellence, a trademark of National Institute for Automotive Service Excellence.

ATA American Trucking Association. Sets standards and influences regulations in the North American truck industry.

Atmospheric Pressure The weight of the air at sea level; 14.696 pounds per square inch (psi) or 101.33 kilopascals (kPa).

Automatic Slack Adjuster The device that automatically adjusts the clearance between the brake linings and the brake drum or rotor. The slack adjuster controls the clearance by sensing the length of the stroke of the pushrod for the air brake chamber. Connects an air brake chamber actuation rod to an S-camshaft.

Auxiliary Filter A device installed in the oil return line between the oil cooler and the transmission to prevent debris from being flushed into the transmission causing a failure. An auxiliary filter must be installed before the vehicle is placed back in service.

Axis of Rotation The centerline around which a gear or part revolves.

Axle (1) A rod or bar on which wheels turn. (2) A shaft that transmits driving torque to the wheels.

Axle Range Interlock A feature designed to prevent axle shifting when the interaxle differential is locked out, or when lockout is engaged. The basic shift system operates the same as the standard shift system to shift the axle and engage or disengage the lockout.

Beam Suspension A tandem suspension relying on a pivotally mounted beam, with axles attached at ends for lead equalization. Beam is mounted to center spring.

Bias A tire term where belts and plies are laid diagonally or crisscrossing each other.

Blink Codes A means of displaying vehicle fault codes without the use of an EST; codes are blinked or flashed using dash display lamps. Also known as flash codes.

Blower Fan A fan that pushes or blows air through a ventilation, heater, or air conditioning system.

Bobtail Proportioning Valve A valve that senses when the tractor is bobtailing and automatically reduces the amount of air pressure that can be applied to the tractor's drive axle(s).

Brake Chamber Air brake application device in which the potential energy of compressed air is converted to mechanical linear force.

Brake Control Valve A dual brake valve that releases air from the service reservoirs to the service lines and brake chambers.

Brake Disc A steel disc used in a braking system with a caliper and pads.

Brake Drum A cast metal bell-like cylinder attached to the wheel that is used to house the brake shoes and provide a friction surface for stopping a vehicle.

Brake Fade A condition that occurs when friction surfaces become hot enough to cause the coefficient of friction to drop to a point where the application of severe pedal pressure results in little actual braking.

Brake Lining Friction material used to line brake shoes or brake pads.

Brake Pad The friction lining and plate assembly that is forced against the rotor to cause braking action in a disc brake system.

British Thermal Unit (Btu) A measure of heat quantity equal to the amount of heat required to raise 1 pound of water 1 °F.

Broken-Back Driveshaft A term often used for nonparallel driveshaft.

Caliper A disc brake component that changes hydraulic pressure into mechanical force and uses that force to press the brake pads against the rotor and stop the vehicle.

Cam Brakes Brake shoes that are actuated by an S-type camshaft that forces them into a drum.

Camber The attitude of a wheel and tire assembly when viewed from the front of a car. If it leans outward, away from the car at the top, the wheel is said to have positive camber. If it leans inward, it is said to have negative camber.

Caster The angle formed between the kingpin axis and a vertical axis as viewed from the side of the vehicle. Caster is considered positive when the top of the kingpin axis is behind the vertical axis.

Cavitation Erosion caused by vapor-bubble collapse.

Center Support Bearing A hanger bearing used to suspend the first section of driveshaft in a double driveshaft assembly

Charging Circuit The alternator (or generator) and associated circuit used to keep the battery charged and to furnish power to the vehicle's electrical systems when the engine is running.

Charging System A system consisting of the battery, alternator, voltage regulator, associated wiring, and the electrical loads of a vehicle.

Check Valve A valve that allows air to flow in one direction only. It is a federal requirement to have a check valve between the wet and dry air tanks.

Chlorofluorocarbon (CFC) A compound used in older A/C refrigerant that damages the ozone layer.

Circuit The complete path of an electrical current, including the generating device. When the path is unbroken, the circuit is closed and current flows. When the circuit continuity is broken, the circuit is open and current flow stops.

Circuit Breaker Circuit protection device that trips when a specific current value is exceeded.

Clean Air Act (CAA) Federal regulations, passed in 1992 that have resulted in major changes in air conditioning systems.

Climbing A gear problem caused by excessive wear in gears, bearings, and shafts whereby the gears move sufficiently apart to cause the apex (or point) of the teeth on one gear to climb over the apex of the teeth on another gear with which it is meshed.

Cloud Point The temperature at which some of the fractions in a diesel fuel begin to wax (solidify) and the fuel takes on an opaque appearance.

Clutch Brake A circular disc with outboard friction surfaces mounted on the transmission input shaft between the release bearing and the transmission.

Coefficient of Friction A measurement of the amount of friction developed between two objects in physical contact when one of the objects is drawn across the other.

Cold Cranking Amperes (CCA) Rating The primary method of rating a vehicle battery. It specifies the current load a battery is capable of delivering for 30 seconds.

Commercial Vehicle Safety Alliance (CVSA) Defines out-of-service criteria used by vehicle safety enforcement officers across North America (United States, Mexico, and Canada) to standardize truck safety standards.

Compression Applying pressure to a spring or any springy substance, thus causing it to reduce its length in the direction of the compressing force.

Condenser A component in an air conditioning system used to cool a refrigerant below its boiling point, causing it to change from vapor to a liquid.

Conductor Any material that permits the electrical current to flow.

Constant-Rate Springs Leaf-type spring assemblies that have a constant rate of deflection.

Continuity Used to describe the status of an electrical connection or circuit with little or no resistance to current flow.

Control Arm The main link between the vehicle's frame and the wheels that acts as a hinge to allow wheel action up and down independent of the chassis.

Controlled Traction A type of differential that uses a friction plate assembly to transfer drive torque from the vehicle's slipping wheel to the one wheel that has good traction or surface bite.

Coolant Liquid that circulates in an engine cooling system.

Crankcase The housing within which the crankshaft and many other parts of the engine operate.

Cross-Tube A system that transfers the steering motion to the opposite, passenger side steering knuckle. It links the two steering knuckles together and forces them to act in unison.

Cycling (1) Repeated on-off action of the air conditioner compressor. (2) Repeated electrical discharge/recharge cycling that can cause the battery positive plate material to break away from its grids and fall into the sediment chambers at the base of the battery case.

Dampen To slow or reduce oscillations or movement.

Dampened Discs Clutch discs that have dampening springs incorporated into the disc hub. When engine torque is first transmitted to the disc, the plate rotates on the hub, compressing the springs. This action absorbs the shocks and torsional vibration caused by today's low-RPM, high-torque engines.

Dash Control Valves A variety of hand-operated valves located on the dash. They include parking brake valves, tractor-protection valves, and differential lock.

Dash Digital Display (DDD) Any of a number of dash-located, digital display units.

Data Links Circuits through which computers communicate with other electronic devices such as control panels, modules, some sensors, or other computers in the form of digital signals.

Dead Axle Nonlive or dead axles are often mounted in lifting suspensions.

Deadline To take a vehicle out of service.

Department of Transportation (DOT) A government agency that establishes vehicle standards.

Diagnostic Flow Chart A chart that provides a systematic approach to the electrical system and component troubleshooting and repair.

Dial Caliper A measuring instrument capable of taking inside, outside, depth, and step measurements.

Differential A gear assembly that transmits power from the driveshaft to the wheels and allows two opposite wheels to turn at different speeds for cornering and traction.

Differential Lock A toggle or push-pull type air switch that locks together the rear axles of a tractor so they pull as one for off-the-road operation.

Differential-Carrier Assembly An assembly that controls the drive axle operation.

Digital Binary Signal A signal that has only two values; on and off.

Digital Diagnostic Reader (DDR) Usually refers to electronic service tools (ESTs) used to access a truck chassis databus.

Digital Volt/Ohmmeter (DVOM) A type of test meter recommended by most manufacturers for use on solid state circuits.

Diode The simplest semiconductor device formed by joining P-type semiconductor material with N-type semiconductor material. A diode allows current to flow in one direction, but not in the opposite direction.

Disc Brake A steel rotor used in a braking system with a caliper and pads.

Dolly Trailer landing gear or caster wheel hoist device.

Double Check Valve Used when a component or circuit must receive or be controlled by the higher of two possible sources of air pressure.

Double Reduction Axle An axle that uses two gear sets for greater overall gear reduction and peak torque development. This design is favored for severe service applications, such as dump trucks, cement mixers, and other heavy haulers.

Drag Link A connecting rod or link between the steering gear, pitman arm, and the steering linkage.

Drive or Driving Gear A gear that drives another gear or causes another gear to turn.

Driveshaft An assembly of one or two universal joints connected to a shaft or tube; used to transmit power from the transmission to the differential.

Driveline Angle The alignment of the transmission output shaft, driveshaft, and rear-axle pinion centerline.

Driver-Controlled Main Differential Lock A type of axle assembly which has greater flexibility over the standard type of single reduction axle because it provides equal amounts of driveline torque to each driving wheel, regardless of changing road conditions. This design also provides the necessary differential action to the road wheels when the truck is turning a corner.

Drum Brake A type of brake system in which stopping friction is created by the shoes pressing against the inside of the rotating drum.

Electronic Control Module (ECM) The module that contains the processing and out driver capability of an electronically controlled system. Depending on the OEM, can be synonymous with ECU; Some OEM's use the term ECM to refer to the engine controller and ECU to refer to other chassis system controllers.

Electronic Control Unit (ECU) The brain of the vehicle.

Electronic Leak Detector Handheld devices with a flexible probe used to seek out refrigerant leaks.

Electronic Service Tool (EST) Any of a number of instruments used to work on vehicle electronics. Can be used to refer to a DMM but more commonly used to describe the instruments that connect to the vehicle databus.

Electronically Programmable Memory (EPROM)
Computer memory that permits adaptation of the ECU to various standard mechanically controlled functions.

Electronics The technology of controlling electricity.

Electrons Negatively charged particles orbiting around every nucleus.

Elliot Axle A solid bar front axle on which the ends span the steering knuckle.

End Yoke The component connected to the output shaft of the transmission to transfer engine torque to the driveshaft.

Engine Brake Any number of devices that convert the role of the vehicle engine into that of a power absorbing retarding mechanism.

EPA Acronym for the Environmental Protection Agency.

Equalizer A suspension device used to transfer and maintain equal load distribution between two or more axles of a suspension. Formerly called a rocker beam.

Equalizer Bracket A bracket for mounting the equalizer beam of a multiple axle spring suspension to a truck or trailer frame while allowing for the beam's pivotal movement. Normally there are three basic types: flange-mount, straddle-mount, and under-mount or side-mount.

Evaporator A component in an air conditioning system used to remove heat from the air passing through it.

Exhaust Brake A sliding restriction valve in the exhaust system located between the manifold and the muffler. A slide mechanism that restricts the exhaust flow, causing exhaust back pressure to build up in the engine's cylinders and effect retarding.

Extra Capacity A term that generally refers to (1) A coupling device that has strength capability greater than standard. (2) An oversized tank or reservoir for a fluid or vapor.

False Brinelling The polishing of a surface that is not damaged.

Fanning the Brakes Applying and releasing the brakes in rapid succession on a long downgrade.

Fault Code A code that is recorded into the computer's memory. A fault code can be read by plugging a special break-out box tester into the computer.

FHWA Acronym for Federal Highway Administration.

Fifth Wheel A coupling device mounted on a truck and used to connect a semitrailer. It acts as a hinge point to allow changes in direction of travel between the tractor and the semitrailer.

Fifth-Wheel Height The distance from the ground to the top of the fifth wheel when it is level and parallel with the ground. It can also refer to the height from the tractor frame to the top

of the fifth wheel. The latter definition applies to data given in fifth wheel literature.

Final Drive The last reduction gear set of a truck.

Fixed Value Resistor An electrical device that is designed to have only one resistance rating, which should not change, for controlling voltage.

Flammable Any material that will easily catch fire or explode.

Flare To spread gradually outward in a bell shape.

Flash Codes A means of displaying vehicle fault codes without the use of an EST; codes are blinked or flashed using dash display lamps. Also known as blink codes.

Flexplate A component used to mount the torque converter to the crankshaft. The flexplate is positioned between the engine crankshaft and the T/C. The purpose of the flexplate is to transfer crankshaft rotation to the shell of the torque converter assembly.

FMVSS No 121 Acronym for Federal Motor Vehicle Safety Standard No 121.

Foot Valve A foot-operated brake valve that controls air pressure to the service chambers.

Foot-Pound An English unit of measurement for torque. One foot-pound is the torque obtained by a force of one pound applied to a foot long wrench handle.

Frame Width The measurement across the outside of the frame rails of a tractor, truck, or trailer.

Fretting A result of vibration that the bearing outer race can pick up from the machining pattern.

Front-Axle Limiting Valve A valve that reduces pressure to the front service chambers, thus eliminating front wheel lockup on wet or icy pavements.

Fully-Oscillating Fifth Wheel A fifth wheel with universal articulation.

Fuse Link A short length of smaller gauge wire installed in a conductor, usually close to the power source.

Fusible Link A term often used for fuse link.

GCW Acronym for gross combination weight.

Gear A disk-like wheel with external or internal teeth that serves to transmit or change motion.

Gear Pitch The number of teeth per given unit of pitch diameter; an important factor in gear design and operation.

Gladhand The connectors between tractor and trailer air lines.

Ground The negatively charged side of a circuit. A ground can be a wire, the negative side of the battery, or the vehicle chassis.

Grounded Circuit A shorted circuit that causes current to return to the battery before it has reached its intended destination.

GVW Acronym for gross vehicle weight.

Halogen Light A lamp having a small quartz/glass bulb that contains a fuel filament surrounded by halogen gas. It is contained within a larger metal reflector and lens element.

Hand Valve (1) A valve mounted on the steering column or dash, used by the driver to apply the trailer brakes independently of the tractor brakes. (2) A hand-operated valve used to control the flow of fluid or vapor.

Heads-Up Display (HUD) A technology used in some vehicles that superimposes data on the driver's normal field of vision. The operator can view the information, which appears to "float" just above the hood at a range near the front of a conventional tractor or truck. This allows the driver to monitor conditions such as limited road speed without interrupting his normal view of traffic.

Heat Exchanger A device used to transfer heat, such as a radiator or condenser.

Heater Control Valve A valve that controls the flow of coolant into the heater core from the engine.

Heavy-Duty Truck A truck that has a GVW of 26,001 pounds or more.

Height Control Valve A simple lever-actuated valve used on an air suspension system to automatically maintain chassis ride height at a preset dimension.

High-Resistance Circuits Circuits that have an increase in circuit resistance, with a corresponding decrease in current.

Hydraulic Brakes Brakes that are actuated by a hydraulic system.

Hydrometer A tester designed to measure the specific gravity of a liquid.

I-Beam Axle An axle designed to give great strength at reasonable weight. The cross section of the axle resembles the letter "I."

In-Line Fuse A fuse that is in series with the circuit in a small plastic fuse holder, not in the fuse box or panel. It is used, when necessary, as a protection device for a portion of the circuit even though the entire circuit may be protected by a fuse in the fuse box or panel.

In-Phase The in-line relationship between the forward coupling shaft yoke and the driveshaft slip yoke of a two-piece driveline.

Insulator A material, such as rubber or glass, that offers high electrical resistance.

Integrated Circuit A component containing diodes, transistors, resistors, capacitors, and other electronic components mounted on a single piece of material and capable of performing numerous functions.

Jake Brake A slang term for Jacobs engine brake.

Kinetic Energy Energy in motion.

Kingpin (1) The pin mounted through the center of the trailer upper coupler (bolster plate) that mates with the fifth wheel locks, securing the trailer to the fifth wheel. The configuration is controlled by industry standards. (2) A pin or shaft on which the steering spindle rotates.

Lateral Runout The wobble or side-to-side movement of a rotating wheel or of a rotating wheel and tire assembly.

Limited-Slip Differential A differential that utilizes a clutch device to deliver power to either rear wheel when the opposite wheel is spinning.

Load Proportioning Valve (LPV) A valve used to redistribute application pressure to front and rear brakes based on vehicle loads. This is a load- or height-sensing valve that senses the vehicle load and proportions the braking between front and rear brakes.

Lockup Torque Converter A torque converter that eliminates the 10-percent slip that takes place between the impeller and turbine at the coupling stage of operation.

Maintenance-Free Battery A battery that does not require the addition of water during normal service life.

Manual Slide Release The release mechanism for a sliding fifth wheel, which is operated by hand.

Master Cylinder In a hydraulic brake system, the dual circuit application control valve.

Metering Valve A valve used on vehicles equipped with front disc and rear drum brakes. It improves braking balance during light brake applications by preventing application of the front disc brakes until pressure is built up in the hydraulic system.

Multiple-Disc Clutch A clutch having a large drum-shaped housing that can be either a separate casting or part of the existing transmission housing.

NATEF Acronym for National Automotive Technicians Education Foundation.

National Institute for Automotive Service Excellence (ASE) A nonprofit organization that has an established certification program for automotive, heavy-duty truck, auto body repair, engine machine shop technicians, and parts specialists.

NHTSA Acronym for National Highway Traffic Safety Administration.

NIOSH Acronym for National Institute for Occupation Safety and Health.

NLGI Acronym for National Lubricating Grease Institute.

Nonlive Axle Nonlive or dead axles are often mounted in lifting suspensions. They hold the axle off the road when the vehicle is traveling empty, and put it on the road when a load is being carried. They are also used as air suspension third axles on heavy straight trucks and are used extensively in eastern states with high axle weight laws.

Nonparallel Driveshaft A type of driveshaft installation whereby the working angles of the joints of a given shaft are equal; however the companion flanges and/or yokes are not parallel.

OEM Acronym for original equipment manufacturer.

Off-Road Any terrain not considered part of the highway system.

Ohm A unit of measured electrical resistance.

Ohm's Law The basic law of electricity stating that in any electrical circuit, current, resistance, and voltage work together in a mathematical relationship.

OOS Acronym for out-of-service (deadlined).

Open Circuit An electrical circuit whose path has been interrupted or broken either accidentally (a broken wire) or intentionally (a switch turned off).

Operational Control Valve A valve used to control the flow of compressed air through the brake system.

Oscillation The rotational movement in either fore/aft or side-to-side direction about a pivot point. Generally refers to fifth wheel designs in which fore/aft and side-to-side articulation are provided.

OSHA Acronym for Occupational Safety and Health Administration.

Out-of-Phase A condition of the universal joint that acts somewhat like one person snapping a rope held by a person at the opposite end. The result is a violent reaction at the opposite end. If both were to snap the rope at the same time, the resulting waves would cancel each other and neither would feel the reaction.

Out-of-Round A wheel or tire defect in which the wheel or tire is not round.

Output Yoke The component that serves as a connecting link, transferring torque from the transmission's output shaft through the vehicle's driveline to the rear axle.

Oval Eccentric or egg-shaped.

Overall Ratio The ratio of the lowest to the highest forward gear in the transmission.

Oxidation Inhibitor (1) An additive used with lubricating oils to keep oil from oxidizing even at very high temperatures. (2) An additive for gasoline to reduce the chemicals in gasoline that react with oxygen.

Pad A disc brake lining and metal back riveted, molded, or bonded together.

Parallel Circuit An electrical circuit that provides two or more paths for the current to flow. Each path has separate resistors and operates independently from the other parallel paths. In a parallel circuit, amperage can flow through more than one resistor at a time.

Parking Brake A mechanically applied brake used to prevent a parked vehicle's movement.

Payload The weight of the cargo carried by a truck, not including the weight of the body.

Pintle Hook The standard coupling mechanism used in truck full-trailers. The function of a pintle hook is to engage to the drawbar eye, which itself is bolted to the tongue or drawbar assembly of the trailer.

Pitman Arm A steering linkage component that connects the steering gear to the linkage at the left end of the center link.

Pitting Surface irregularities resulting from corrosion.

Planetary Drive A planetary gear reduction set where the sun gear is the drive and the planetary carrier is the output.

Planetary Gear Set A system of gearing that is somewhat like the solar system. A pinion is surrounded by an internal ring gear and planet gears are in mesh between the ring gear and pinion around which all revolve.

Plies The layers of rubber-impregnated fabric that make up the body of a tire.

PM Acronym for preventive maintenance.

PMI Acronym for preventive maintenance inspections. There are three recognized categories, which are: The "A" PM service, which is the simplest, the "B" inspection, which is described as a major inspection, and the "C" inspection, which is a major inspection plus adjustments and minor repairs.

Polarity The particular state, either positive or negative, with reference to the two poles or to electrification.

Pole The number of input circuits made by an electrical switch.

Pounds per Square Inch (psi) A unit of standard measurement for pressure.

Pour Point The temperature at which diesel fuel begins to solidify to the extent it will not flow properly.

Power A measure of work being done factored with time.

Power Steering A steering system utilizing hydraulic pressure to reduce the turning effort required of the operator.

Power Takeoff (PTO) Any of a number of different means of gear driving auxiliary apparatus: A PTO can be mounted to the engine, transmission, transfer case, or elsewhere on the vehicle drivetrain.

Powertrain Power delivery components from the engine through to the wheels.

Pressure The amount of force applied to a defined area measured in pounds per square inch (psi) English or kilopascals (kPa) metric.

Pressure Differential The difference in pressure between any two points of a system or a component.

Pressure Relief Valve (1) A valve located on the wet tank, usually preset at 150 psi (1,034 kPa). Limits system pressure if the compressor or governor unloader valve malfunctions. (2) A valve located on the rear head of an air conditioning compressor or pressure vessel that opens if an excessive system pressure is exceeded.

Primary Circuit In brake systems refers to the lead circuit in the system.

Printed Circuit Board An electronic circuit board made of thin nonconductive plastic-like material onto which conductive metal, such as copper, has been deposited. Parts of the metal are then etched away by an acid, leaving metal lines that form the conductors for the various circuits on the board. A printed circuit board can hold many complex circuits in a very small area.

Programmable Read Only Memory (PROM) An electronic component that contains program information specific to different vehicle model calibrations.

Proportioning Valve A valve used on vehicles equipped with front disc and rear drum brakes. It is installed in the lines to the rear drum brakes, and in a split system, below the pressure differential valve. By reducing pressure to the rear drum brakes, the valve helps to prevent premature lockup during severe brake application and provides better braking balance.

Psi Acronym for pounds per square inch.

Pull Circuit A circuit that brings the cab from a fully tilted position up and over the center.

Pull-Type Clutch A type of clutch that does not push the release bearing toward the engine; instead, it pulls the release bearing toward the transmission.

Push-Type Clutch A type of clutch in which the release bearing is not attached to the clutch cover.

Quick Release Valve A device used to exhaust air as close as possible to the service chambers or spring brakes.

Radial A tire design having cord materials running in a direction from the center point of the tire, usually from bead to bead.

Radial Load A load that is applied at 90 degrees to an axis of rotation.

Random Access Memory (RAM) The memory used during computer operation to store temporary information. The microprocessor can write, read, and erase information from RAM in any order, which is why it is called random.

Range Shift Cylinder A component located in the auxiliary section of the transmission. This component, when directed by air pressure via low and high ports, shifts between high and low range of gears.

Range Shift Lever A lever located on the shift knob allows the driver to select low or high gear range.

Ratio Valve A valve used on the front or steering axle of a heavy-duty truck to limit the brake application pressure to the actuators during normal service braking.

Reactivity The characteristic of a material that causes it to react with air, heat, water, or other materials.

Read Only Memory (ROM) Optically or magnetically retained memory used in microcomputers to store information permanently.

Reference Voltage The voltage supplied to a sensor by a vehicle computer.

Refrigerant A fluid capable of vaporizing at a low temperature used in A/C and reefer systems.

Relay An electric switch that allows a small current to control a much larger one. It consists of a control circuit and a power circuit.

Relay/Quick Release Valve A pneumatic valve used on air-brake equipped trucks. It is used to speed the application and release of air to the service chambers.

Release Bearing A unit used to disengage a clutch that mounts on the transmission input shaft but does not rotate with it.

Reserve Capacity Rating The ability of a battery to sustain a minimum vehicle electrical load in the event of a charging system failure.

Resistance The opposition to current flow in an electrical circuit.

Revolutions per Minute (rpm) The number of complete turns a member makes in one minute.

Right-to-Know Law A law passed by the federal government and administered by the Occupational Safety and Health Administration (OSHA) that requires any company that uses or produces hazardous chemicals or substances to inform its employees, customers, and vendors of any potential hazards that may exist in the workplace as a result of using the products.

Rigid Disc A steel clutch plate to which friction linings, or facings, are bonded or riveted.

Rigid Fifth Wheel A coupling plate that is fixed rigidly to a frame. This fifth wheel has no articulation or oscillation. It is used in applications where the articulation is provided by other means, such as an articulating upper coupler of a frame-less dump.

Ring Gear (1) The cranking gear of a flywheel. (2) A large circular gear such as that found in a final drive assembly.

Roll Axis The theoretical line that joins the roll center of the front and rear axles.

Rotary Oil Flow A condition caused by the centrifugal force applied to the fluid as the converter rotates around its axis.

Rotation A term used to describe the fact that a gear, shaft, or other device is turning.

Rotor (1) A part of the alternator that provides the magnetic fields necessary to create a current flow. (2) The rotating member of an assembly.

Runout A deviation of the specified normal travel of an object. The amount of deviation or wobble a shaft or wheel has as it rotates. Runout is measured with a dial indicator.

S-Cam Brakes Drum brake system in which the shoes are driven into the drum by rotating an S-shaped cam, spreading the shoes.

Screw Pitch Gauge A gauge used to check the threads per inch of a nut or bolt.

Secondary Circuit In brake systems refers to the following circuit in the system, usually actuated by the primary circuit.

Section Height The tread center to bead plane on a tire.

Section Width The measurement on a tire from sidewall to sidewall.

Self-Adjusting Clutch A clutch that automatically takes up the slack between the pressure plate and clutch disc as wear occurs.

Semiconductor A solid state device that can function as either a conductor or an insulator, depending on how its structure is arranged.

Sensing Voltage The voltage that allows the regulator to sense and monitor the battery voltage level.

Sensor An electronic device used to signal vehicle computers.

Series Circuit An electrical circuit with a voltage source and a single path for current flow.

Series/Parallel Circuit A circuit with both series and parallel branches.

Shift Fork The Y-shaped component located between the gears on the main shaft that, when actuated, causes the gears to engage or disengage via the sliding clutches.

Shift Point Adjustment The means of adjusting the shift RPM in automatic transmissions.

Shift Rail Shift rails guide the shift forks using a series of grooves, tension balls, and springs to hold the shift forks in gear.

Shift Tower The main interface between the driver and the transmission, consisting of a gearshift lever, pivot pin, spring, boot and housing.

Shift Yoke A Y-shaped component located between the gears on the main shaft that, when actuated, causes the gears to engage or disengage via the sliding clutches.

Shock Absorber A hydraulic device used to dampen vehicle spring oscillations for controlling body sway and wheel bounce, and/or prevent spring breakage.

Short Circuit An undesirable connection between two worn or damaged wires. The short occurs when the insulation is worn between two adjacent wires and the metal in each wire contacts the other, or when the wires are damaged or pinched.

Single-Reduction Axle Any axle assembly that uses a single gear reduction in the differential carrier.

Sliding Fifth Wheel A fifth wheel design that slides fore and aft to adjust weight distribution on the tractor axles and/or overall length of the tractor and trailer.

Slipout A condition that generally occurs when pulling with full power or decelerating with the load pushing. Tapered or worn clutching teeth will try to "walk" apart as the gears rotate, causing the sliding clutch and gear to slip out of engagement.

Solenoid An electromagnet that is used to effect linear or rotary movement.

Spalling Surface fatigue that occurs when chips, scales, or flakes of metal break off due to fatigue rather than wear. Spalling is usually found on splines and U-joint bearings.

Specific Gravity The ratio of a liquid's mass to an equal volume of distilled water.

Spiral-Bevel Gear A gear arrangement that has a drive pinion gear that meshes with the ring gear at the centerline axis of the ring gear. This gearing provides strength and allows for quiet operation.

Splined Yoke A yoke that allows the driveshaft to increase in length to accommodate movements of the drive axles.

Spread-Tandem Suspension A two-axle assembly in which the axles are spaced to allow maximum axle loads under existing regulations. The distance is usually more than 55 inches.

Spring Rate The load required to deflect the spring a given distance (usually 1 inch).

Spring Spacer A riser block often used on top of the spring seat to obtain increased mounting height.

Starter Motor The device that converts the electrical energy from the battery into mechanical energy for cranking the engine.

Storage Battery A lead-acid battery.

Synchro-Transmission A transmission with mechanisms for synchronizing the gear speeds so that the gears can be shifted without clashing, thus eliminating the need for double-clutching.

Tachometer An instrument that indicates rotating speeds, sometimes used to indicate crankshaft RPM.

Tag Axle A nondrive rearmost axle of a tandem axle tractor used to increase the load-carrying capacity of the vehicle.

Tandem Drive A two-axle drive combination.

Thermostat Any of a number of different types of heat sensing switches.

Throttle Position Sensor (TPS) Usually a potentiometer type sensor using a three-wire voltage divider principle. IT signals mechanical accelerator position to the engine or chassis EMC.

Throw (1) The offset of a crankshaft. (2) The number of output circuits of a switch.

TMC The Maintenance Council of the American Trucking Association. Sets standards for the trucking industry including that for wheel end procedure.

Toe A suspension dimension that reflects the difference in the distance between the extreme front and rear of the tire.

Toe-In A suspension dimension whereby the front of the tire points inward toward the vehicle.

Toe-Out A suspension dimension whereby the front of the tire points outward from the vehicle.

Torque Converter A component device, similar to a fluid coupling, that transfers engine torque to the transmission input shaft and can multiply engine torque by having one or more stators between the members.

Tractor Protection Valve A device that automatically seals off the tractor air supply from the trailer air supply when system pressure drops to 20 or 45 psi.

Trailer Hand-Control Valve A device located on the dash or steering column and used to apply only the trailer brakes.

Treadle Valve A foot-operated brake valve that controls air pressure to the service chambers.

Two-Speed Axle Assembly An axle assembly having two different output ratios from the differential.

U-bolt A fastener used to clamp the top U-bolt plate, spring, axle, and bottom U-bolt plate together.

Universal Joint (U-Joint) A shaft component that allows torque to be transmitted at different angles.

Vehicle Body Clearance (VBC) The distance from the inside of the inner tire to the spring or other body structures.

Vehicle On-board Radar (VORAD) A Doppler-radar collision warning system.

Vehicle Retarder An auxiliary braking device designed to supplement the service brakes on heavy-duty trucks.

Vertical Load Capacity The maximum, recommended vertical downward force that can be safely applied to a coupling device.

VIN Acronym for Vehicle Identification Number.

Viscosity A measure of the flow and sheer resistance of lubricating oil.

Voltage Regulator A device that controls the voltage in the charging circuit.

Wedge-Actuated Brakes A brake system using air pressure and air brake chambers to drive a wedge and roller assembly into an actuator located between brake shoes.

Wet Tank A supply reservoir.

CPSIA information can be obtained
at www.ICGtesting.com
Printed in the USA
BVHW070447101122
651641BV00006B/41